Frontiers in Mathematics

Advisory Editors

Laurent Saloff-Coste, Cornell University, Ithaca, NY, USA
Igor Shparlinski, The University of New South Wales, Sydney, NSW, Australia
Wolfgang Sprößig, TU Bergakademie Freiberg, Freiberg, Germany

This series is designed to be a repository for up-to-date research results which have been prepared for a wider audience. Graduates and postgraduates as well as scientists will benefit from the latest developments at the research frontiers in mathematics and at the "frontiers" between mathematics and other fields like computer science, physics, biology, economics, finance, etc. All volumes are online available at SpringerLink.

Gabriela Ileana Sebe • Dan Lascu

Metrical and Ergodic Theory of Continued Fraction Algorithms

Gabriela Ileana Sebe
Faculty of Applied Sciences
National University of Science and Technology
Politehnica Bucharest
Bucharest, Romania

Dan Lascu
Romanian Naval Academy "Mircea cel Batran"
Constanta, Romania

Gheorghe Mihoc-Caius Iacob
Institute of Mathematical Statistics and Applied
Mathematics of the Romanian Academy
Bucharest, Romania

ISSN 1660-8046
Frontiers in Mathematics
ISBN 978-3-031-86633-3
https://doi.org/10.1007/978-3-031-86634-0

ISSN 1660-8054 (electronic)

ISBN 978-3-031-86634-0 (eBook)

© The Editor(s) (if applicable) and The Author(s), under exclusive license to Springer Nature Switzerland AG 2025

This work is subject to copyright. All rights are solely and exclusively licensed by the Publisher, whether the whole or part of the material is concerned, specifically the rights of translation, reprinting, reuse of illustrations, recitation, broadcasting, reproduction on microfilms or in any other physical way, and transmission or information storage and retrieval, electronic adaptation, computer software, or by similar or dissimilar methodology now known or hereafter developed.

The use of general descriptive names, registered names, trademarks, service marks, etc. in this publication does not imply, even in the absence of a specific statement, that such names are exempt from the relevant protective laws and regulations and therefore free for general use.

The publisher, the authors and the editors are safe to assume that the advice and information in this book are believed to be true and accurate at the date of publication. Neither the publisher nor the authors or the editors give a warranty, expressed or implied, with respect to the material contained herein or for any errors or omissions that may have been made. The publisher remains neutral with regard to jurisdictional claims in published maps and institutional affiliations.

This book is published under the imprint Birkhäuser, www.birkhauser-science.com by the registered company Springer Nature Switzerland AG
The registered company address is: Gewerbestrasse 11, 6330 Cham, Switzerland

If disposing of this product, please recycle the paper.

Preface

This book presents the most important results of our research in the metrical theory of continued fractions in the last 9 years.

Except for the classical theory of regular continued fractions (RCFs) based on the famous Gauss map, a large amount of research has been devoted to the study of various algorithms for the representation of real numbers by means of sequences of integers.

The metrical theory of the continued fraction expansion is about the sequence of its incomplete quotients and related sequences. This theory has connections with many fields, including probability theory, number theory, ergodic theory, and dynamical systems (see, e.g., [14, 20, 24, 34, 47, 48, 56, 70]).

The book is organized in five chapters. The core of the book consists of the systematic study of three large families of continued fractions. To understand the global behavior of such families of expansions of numbers, the properties of dynamical systems that generate them are explored. Since the ergodic results do not yield rates of convergence for mixing properties, Gauss-Kuzmin-type theorems are needed.

Chapter 1 has two purposes. The first one is to give a short presentation of the classical RCF expansion and to describe some historical aspects concerning the Gauss-Kuzmin-Lévy Problem together with its current developments. The second goal is to present some concepts, notations, and general results. We describe the main properties of Perron-Frobenius operators and we discuss the construction of a homogeneous Markov chain with an arbitrary measurable state space. The most important part of this chapter is a survey of random systems with complete connections (RSCCs) and the associated Markov chains (MCs). We focus on a special class of MCs, called compact MCs. An important class of compact MCs with great impact in the metrical theory of continued fractions is that of MCs associated with RSCCs with contraction on a compact metric space. Thus, the existence and uniqueness of a stationary probability measure for the chain is investigated. These results are very useful in order to study the asymptotic behavior of the associated transition operator of the MC on suitable Banach spaces.

Having always in view the classical RCF expansion, we start with θ-continued fraction expansions of a number in terms of an irrational $\theta \in (0, 1)$ in Chap. 2. Their properties are not essentially different from those of the RCF expansion. It is worth noting that these new continued fraction expansions are generated by transformations which are ergodic with respect to an invariant probability measure only in the particular case $\theta^2 = 1/m$, $m \in \mathbb{N}_+$. There are still many open questions closely related to this problem. Since the ergodic properties are not enough to yield rates of convergence for mixing properties, we obtain some Gauss-Kuzmin-type theorems.

Chapter 3 is devoted to our study of the family of N-continued fraction expansions, for integers $N \geq 2$. We use the method of dependence with complete connections to solve its Gauss-Kuzmin-type problem, and we also give a two-dimensional Gauss-Kuzmin theorem.

Chapter 4 considers generalized Rényi CFs, which represent a special class of backward CFs. Using ergodic properties of the random system with complete connections associated with the underlying dynamical system, we prove a Gauss-Kuzmin-type theorem. Next, using Szüsz's method we obtain more information on the convergence rate involved. Finally, using a Wirsing-type approach to the Perron-Frobenius operator of the Renyi-type continued fraction transformation under its invariant measure we study the optimality of the convergence rate.

Since there are several CF algorithms, we ask ourselves which of them provides the best approximation of a real number. The representation of a real number by a continued fraction can be viewed as a source of information about the number. The last chapter presents the investigation of the efficiency of the families of continued fraction studied in the previous chapters. In addition, a new family of continued fractions whose digits are non-positive integer powers of integers $\ell \geq 2$, studied by us in [38, 62] is considered. These expansions can be associated with some random Fibonacci-type sequences.

Using the metric properties of expansions, some results concerning the appropriate Kolmogorov-Sinai entropy in the classification of dynamical systems and an extended result of Lochs [45], we can compare any two expansions of numbers which are generated by surjective transformations, under certain conditions.

As we have already mentioned, all the results presented here are original.

The references list exceeds the number of works quoted in the book. It should be consulted with the purpose of starting new investigations.

Bucharest, Romania Gabriela Ileana Sebe
Constanta, Romania Dan Lascu
December 2024

Declarations

Competing Interests The authors have no competing interests to declare that are relevant to the content of this manuscript.

Contents

1 **Introduction** 1
   1.1 Gauss' Problem and Current Developments 1
   1.2 Perron-Frobenius Operators 3
   1.3 Markov Chains with an Arbitrary State Space 6
   1.4 Random Systems with Complete Connections 7

2 **θ-Continued Fraction Expansions** 13
   2.1 θ-Continued Fraction Expansion as Dynamical System 13
   2.2 Basic Metric Properties 14
   2.3 Some Ergodic Properties 18
   2.4 An Infinite-Order-Chain Representation 20
      2.4.1 Natural Extension of T_θ 20
      2.4.2 Extended Random Variables 21
   2.5 The Perron-Frobenius Operator of T_θ Under γ_θ 23
   2.6 Ergodicity of the Associated RSCC 26
   2.7 A Gauss-Kuzmin-Type Theorem for T_θ 28
   2.8 Szüsz's Method to Gauss-Kuzmin-Type Theorem 29
   2.9 A Near-Optimal Solution to the Gauss-Kuzmin-Lévy Problem 33
      2.9.1 Appendix A1 38
      2.9.2 Appendix A2 39
   2.10 Gauss-Kuzmin Theorem Related to the Natural Extension 40
   2.11 Two Asymptotic Distributions 45

3 **N-Continued Fractions** 51
   3.1 Preliminary Considerations 51
   3.2 Basic Metric Properties 56
   3.3 The Natural Extension of T_N and Extended Random Variables 58
      3.3.1 Natural Extension of T_N 58
      3.3.2 Extended Random Variables 59
   3.4 Perron-Frobenius Operators 62
   3.5 Gauss-Kuzmin Theorem for N-Continued Fractions 66
      3.5.1 The One-Dimensional Case 66

		3.5.2	A Two-Dimensional Case	71
		3.5.3	Lower and Upper Bounds for the Convergence Rate of the Distribution Function	76

4 Generalized Rényi Continued Fractions ... 85
4.1 Introduction ... 85
4.2 Rényi-Type Continued Fraction Expansion as Dynamical System ... 87
4.3 The Probabilistic Structure of $(r_n)_{n\in\mathbb{N}_+}$ under the Lebesgue Measure ... 88
4.4 Natural Extension and Extended Random Variables ... 91
4.4.1 Natural Extension of R_N ... 91
4.4.2 Extended Random Variables ... 92
4.5 Perron-Frobenius Operators ... 94
4.6 The Gauss-Kuzmin-Type Theorem for the Rényi-Type Continued Fraction Expansions ... 99
4.7 Szüsz's Method to Gauss-Kuzmin-Lévy-Type Theorem ... 102
4.8 Wirsing-Type Approach to the Perron-Frobenius Operator ... 108
4.8.1 Near-Optimal Solution to the Gauss-Kuzmin-Lévy Problem ... 108
4.8.2 Final Remarks ... 113
4.9 A Two-Dimensional Gauss-Kuzmin Theorem ... 114

5 Comparing the Efficiency of Different Continued Fraction Algorithms ... 123
5.1 Entropy of Dynamical Systems ... 123
5.2 A Lochs-Type Approach ... 125
5.3 Dajani-Fieldsteel Theorem ... 127
5.4 Applications of the Rohlin's Entropy Formula ... 127
5.4.1 θ-Continued Fraction Expansions ... 128
5.4.2 Chan's Continued Fraction Expansions ... 128
5.4.3 N-Continued Fraction Expansions ... 131
5.4.4 Rényi-Type Continued Fraction Expansions ... 132
5.5 Comparing the Efficiency of Some Expansions ... 133
5.5.1 N-Continued Fractions and Chan's Continued Fractions ... 133
5.5.2 N-Continued Fractions and Rényi-Type Continued Fractions ... 134
5.5.3 Rényi-Type Continued Fractions and Chan's Continued Fractions ... 134
5.6 Final Remarks ... 135

Bibliography ... 137

List of Notations

Abbreviations

a.e.	Almost everywhere (with respect to Lebesgue measure)
a.s.	Almost surely (with respect to any other measure)
γ	Gauss' measure
log	Natural logarithm
CF	Continued fraction
MC	Markov chain
RCF	Regular continued fraction
RSCC	Random system with complete connections
supp	Support of a measure
var	Total variation
□	End of proof

Symbols

$\mathbb{N}$	$\{0, 1, 2, \ldots\}$, $\mathbb{N}_+ = \{1, 2, \ldots\}$
$\mathbb{Z}$	$\{\ldots, -1, 0, 1, 2, \ldots\}$
$\mathbb{Q}$	The set of rational numbers
$\mathbb{R}$	The set of real numbers
$\mathbb{R}_+$	The set of non-negative real numbers
$\mathbb{C}$	The set of complex numbers
I	$[0, 1]$ The unit interval of $\mathbb{R}$
$\Omega = I \setminus \mathbb{Q}$	The set of irrationals in I
$\lfloor a \rfloor$	Integer part of $a \in \mathbb{R}$
$\mathcal{B}_I$	σ-algebra of Borel sets in I
$\mathcal{B}_I^2$	σ-algebra of Borel sets in I^2
E	Expected value (mean)

$\mathcal{P}(A)$	The power set of A
χ_A	Characteristic function of the set A, i.e., $\chi_A = \begin{cases} 1, \text{ if } x \in A \\ 0, \text{ if } x \notin A \end{cases}$
λ	Lebesgue measure on $\mathcal{B}_I$
$\ll$	Absolute continuity of measures
$\otimes$	Product of σ-algebras or measures
g	$(\sqrt{5}-1)/2 = 0.6180339887\ldots$
g^2	$(3-\sqrt{5})/2 = 0.38196\ldots$ and $g + g^2 = 1$
λ_0	$0.303663002898732658\ldots$ (Wirsing's constant)
$\mathrm{Li}_2(z)$	$\sum_{k \geq 1} \dfrac{z^k}{k^2}$ or $\mathrm{Li}_2(z) = \int_z^0 \dfrac{\ln(1-t)}{t}dt$-dilogarithm function
$\zeta(2)$	$\sum_{i \geq 1} i^{-2} = \pi^2/6$-Riemann zeta function
$\zeta(2, N)$	$\sum_{i \geq N} i^{-2}$-Hurwitz zeta function
$\zeta(3, N)$	$\sum_{i \geq N} i^{-3}$-Hurwitz zeta function

Introduction

1.1 Gauss' Problem and Current Developments

The metrical theory of continued fractions started on 25th October 1800 with a note by C.F. Gauss in his mathematical diary. Gauss wrote that, in modern notation,

$$\lim_{n \to \infty} \lambda \left(\tau^n \leq x \right) = \frac{\log(1+x)}{\log 2}, \quad x \in I := [0,1]. \tag{1.1.1}$$

Here λ is the Lebesgue measure and $\tau : [0, 1) \to [0, 1)$ is the Gauss map defined by

$$\tau(x) := \frac{1}{x} - \left\lfloor \frac{1}{x} \right\rfloor, \, x \neq 0; \, \tau(0) = 0, \tag{1.1.2}$$

where $\lfloor \cdot \rfloor$ stands for integer part. For each $x \in (0, 1)$ put $a_n = a_n(x) := a_1\left(\tau^{n-1}(x)\right)$, $n \in \mathbb{N}_+ := \{1, 2, \ldots\}$ with $\tau^0(x) = x$ and $a_1 = a_1(x) := \lfloor 1/x \rfloor$, $x \neq 0$. Denote by Ω the set of irrationals in the unit interval I. Any irrational $\omega \in \Omega$ can be written as the infinite regular continued fraction (RCF)

$$\omega = \cfrac{1}{a_1(\omega) + \cfrac{1}{a_2(\omega) + \cfrac{1}{a_3(\omega) + \cfrac{}{\ddots}}}} =: [a_1(\omega), a_2(\omega), a_3(\omega), \ldots]_G. \tag{1.1.3}$$

The positive integers $a_n(\omega)$, $n \in \mathbb{N}_+$, are called the (RCF) *digits* (also known as *partial quotients* or *incomplete quotients*) of $\omega \in \Omega$. Letting $\mathcal{B}_I$ denote the σ-algebra of Borel subsets of I, it is obvious that the digits a_n, $n \in \mathbb{N}_+$, are positive integer-valued random

variables which are defined almost surely on $(I, \mathcal{B}_I)$ with respect to any probability measure on $\mathcal{B}_I$, which assigns probability zero to the set $I \setminus \Omega$ of rationals in I. An example of such a measure is Lebesgue measure λ, but a more important one is *Gauss' measure*
$$\gamma([0,x]) := \frac{\log(1+x)}{\log 2}.$$

Gauss' proof has never been found. In 1812 Gauss asked Laplace for an estimate of the n-th error term
$$e_n(x) = \lambda\left(\tau^n \leq x\right) - \frac{\log(1+x)}{\log 2}, \quad n \in \mathbb{N}_+, x \in I. \tag{1.1.4}$$

This has been called *Gauss' Problem*. The first one to give a solution to Gauss' Problem, implicitly proving Gauss' 1800 assertion, was R.O. Kuzmin, who showed in 1928 that $e_n(x) = \mathcal{O}(q^{\sqrt{n}})$ as $n \to \infty$, uniformly in x with some unspecified $0 < q < 1$. Kuzmin's proof is reproduced in Khintchine's book [35].

Independently, one year later, P. Lévy [46] improved Kuzmin's result by showing that $|e_n(x)| \leq q^n$ for $n \in \mathbb{N}_+, x \in I$, with $q = 3.5 - 2\sqrt{2} = 0.67157\ldots$. The Gauss-Kuzmin-Lévy theorem is the first basic result in the rich metrical theory of continued fractions.

Since in the results of Kuzmin and Lévy the constants involved are far from optimal, in the sixty years that followed, a "hunt" for the best possible constant started.

Using Kuzmin's approach in 1961 P. Szüsz [73] claimed to have lowered the Lévy estimate for q to 0.4. Actually, Szüsz's argument yields just 0.483 rather than 0.4.

In 1974 a decisive step in the final solution of Gauss' Problem was taken by E. Wirsing [76] who found that the optimal value of q is $0.303663002898732658\ldots$ (Wirsing's constant).

This is not the whole story of the Gauss-Kuzmin-Lévy theorem. In 1978 K.I. Babenko [2] obtained what might be called the complete solution of Gauss' Problem, namely to express $\lambda(\tau^n \leq x)$ in terms of the eigenvalues and eigenfunctions of a linear operator related to the transition operator of the random system with complete connections associated with the RCF expansion.

Ramifications of this problem for RCF are extensively studied in [27, 29]. An important aspect in the metrical theory of the RCF expansion is that Gauss' measure γ represents the invariant measure of the map τ underlying the RCF, i.e., $\gamma = \gamma \tau^{-1}$. Another aspect is that the measure preserving dynamical system $(I, \mathcal{B}_I, \gamma, \tau)$ is ergodic.

Besides γ, the probability measures on $\mathcal{B}_I$, $(\gamma_a)_{a \in I}$, defined by their distribution functions
$$\gamma_a([0,x]) := \frac{(a+1)x}{ax+1}, \quad x \in I, a \in I,$$

which we call *conditional*, are the most natural ones associated with the RCF expansions. In particular, $\gamma_0 = \lambda$. This idea goes back to W. Doeblin [18] who was the first to use dependence with complete connections in the metrical theory of the RCF expansions.

1.2 Perron-Frobenius Operators

We may synthesize one of the most relevant results for RCFs as follows [29, pp. 117–119, p. 154 and p. 98].

For any $a, x \in I$ and $n \in \mathbb{N}_+$ we have

$$\left| \gamma_a \left(\tau^n \leq x \right) - \frac{\log(x+1)}{\log 2} \right| \leq \left(\frac{\pi^2 \log 2}{6} - 1 \right) \lambda_0^{n-1} \left(\frac{1}{2} - \left| \frac{1}{2} - \frac{\log(x+1)}{\log 2} \right| \right),$$

where λ_0 is Wirsing's constant. For any $a \in I$, $a \neq a_0$, with a_0 very close to 0.4 the exact convergence rate to 0 as $n \to \infty$ of

$$\sup_{x \in I} \left| \gamma_a \left(\tau^n \leq x \right) - \frac{\log(x+1)}{\log 2} \right|$$

is $\mathcal{O}\left(\lambda_0^n \right)$. For $a = a_0$ the rate is $\mathcal{O}\left(\lambda_3^n \right)$ with $\lambda_3 = 0.100884509293\ldots$. We thus see that there exists a case where even faster convergence than Wirsing's does hold. Note that these results needed nearly 162 years to be reached.

New problems were raised whose solutions do not seem easy. A detailed treatment of Gauss-Kuzmin-Lévy problem is based on functional-theoretic methods. These are applied to the Perron-Frobenius operator associated with τ, considered as acting on various Banach spaces. The use of the ergodicity of τ under γ for proving ergodic properties of the RCF expansion has been initiated by W. Doeblin [18] in 1940, 11 years before C. Ryll-Nardzewski [56], who is widely regarded as the first having used ergodic theorems in the metrical theory of RCFs.

In Sect. 1.2 we collect elements of ergodic theory and Perron-Frobenius operators. For our further analysis we shall need some basic notions of Markov chains (see Sect. 1.3) and some general concepts regarding random systems with complete connections (see Sect. 1.4).

Generalizations of these problems for non-RCFs are also known as the Gauss-Kuzmin-Lévy problem. The remarkable fact is that in the last 25 years, such an investigation has been made in the two-dimensional case. The explorations of the two-dimensional Gauss-Kuzmin theorem for other continued fractions were fulfilled in [15, 58, 59, 66].

1.2 Perron-Frobenius Operators

Let $(X, \mathcal{X}, \mu)$ be a measure space. A transformation T of X is said to be *measure preserving* with respect to μ (equivalently, the measure μ is T-invariant) if $\mu\left(T^{-1}(A)\right) = \mu(A)$ for every μ-measurable set $A \in \mathcal{X}$. If T is invertible, the last equation is equivalent to $\mu(T(A)) = \mu(A)$ for every μ-measurable set $A \in \mathcal{X}$. A transformation T of X is said to be μ-*non-singular* if $\mu(T^{-1}(A)) = 0$ for all $A \in \mathcal{X}$ for which $\mu(A) = 0$. Clearly, any μ-preserving transformation is μ-non-singular. Let ν and μ be two measures on the measure space $(X, \mathcal{X})$. We say that ν is *absolutely continuous* with respect to μ if for any

$A \in \mathcal{X}$, such that $\mu(A) = 0$, it follows that $\nu(A) = 0$. We write $\nu \ll \mu$. If $\nu \ll \mu$, then we may represent ν in terms of μ. Let $L^1_\mu(X) := \{f : X \to \mathbb{C} | \int_X |f| d\mu < \infty\}$.

Theorem 1.2.1 (Radon-Nikodym Theorem) *Let $(X, \mathcal{X})$ be a measurable space and let ν and μ two probability measures on $(X, \mathcal{X})$. If $\nu \ll \mu$, then there exists a unique $f \in L^1_\mu(X)$ such that for every $A \in \mathcal{X}$,*

$$\nu(A) = \int_A f d\mu. \tag{1.2.1}$$

The function f is called the Radon-Nikodym derivative and is denoted by $\dfrac{d\nu}{d\mu}$.

A *dynamical system* is a quadruple $(X, \mathcal{X}, \mu, T)$, where $(X, \mathcal{X}, \mu)$ is a probability space and $T : X \to X$ is a surjective μ-preserving transformation. If $(X, \mathcal{X}, \mu, T)$ is a dynamical system, then T is called *ergodic* if for every μ-measurable set A satisfying $T^{-1}(A) = A$ one has $\mu(A) = 0$ or $\mu(A) = 1$. Such a set A is called *T-invariant*.

Let $(X, \mathcal{X}, \mu)$ be a probability space. For a μ-non-singular transformation T, the *Perron-Frobenius operator* (or, *transfer operator [55]*) P_μ associated with T is defined as the bounded linear operator on the Banach space $L^1_\mu(X)$ which satisfies

$$\int_A P_\mu f \, d\mu = \int_{T^{-1}(A)} f \, d\mu \quad \text{for all } f \in L^1_\mu(X), A \in \mathcal{X}. \tag{1.2.2}$$

The existence and uniqueness of $P_\mu f$ follows from the Radon-Nikodym theorem. This also implies the existence and uniqueness of P_μ. Furthermore, the following holds:

Proposition 1.2.2

(i) *P_μ is positive, that is, $P_\mu f \geq 0$ if $f \geq 0$ ([6], p. 78).*
(ii) *P_μ preserves integrals, that is, $\int_X P_\mu f \, d\mu = \int_X f \, d\mu$ for all $f \in L^1_\mu(X)$ ([6], p. 78).*
(iii) *For all $n \in \mathbb{N}_+$, the n-th power $(P_\mu)^n$ of P_μ is the Perron-Frobenius operator associated with the n-th iterate T^n of T.*
(iv) *There exists $f \in L^1_\mu(X)$ such that $f \geq 0$ and $P_\mu f = f$ a.e. if and only if T preserves the measure ν which is defined as $\nu(A) := \int_A f d\mu$ for $A \in \mathcal{X}$. In particular, $P_\mu 1 = 1$ if and only if T is μ-preserving ([6], p. 80).*

The probabilistic interpretation of P_μ is immediate: If an X-valued random variable ξ on X has μ-density h, that is, $\mu(\xi \in A) = \int_A h d\mu$, $A \in \mathcal{X}$, with $h \geq 0$ and $\int_X h d\mu = 1$, then $T \circ \xi$ has μ-density $P_\mu h$.

1.2 Perron-Frobenius Operators

In the special case $\mu = \lambda$, we obviously have

$$P_\lambda f(x) = \frac{d}{dx} \int_{T^{-1}([0,x])} f \, d\lambda \quad \text{a.e. in } X. \tag{1.2.3}$$

The asymptotic properties of the Perron-Frobenius operator $P_\mu : L^1_\mu \to L^1_\mu$ are not strong enough for to lead to a satisfactory solution to Gauss' problem, while when restricting to other Banach spaces they are substantially better. In the next chapters the domain of P_μ will be successively restricted to various Banach spaces. Therefore, in this section we describe some Banach spaces which are often mentioned.

Let J be a finite real interval. We denote by $B(J)$ the collection of all bounded measurable functions $f : J \to \mathbb{C}$. This is a commutative Banach algebra with unit under the norm

$$\|f\| := \sup_{x \in J} |f(x)|, \quad f \in B(J).$$

We denote by $C(J)$ the collection of all continuous functions $f : J \to \mathbb{C}$. This is a Banach space under the supremum norm. We denote by $C^1(J)$ the collection of all functions $f : J \to \mathbb{C}$ which have a continuous derivative. This is a commutative Banach algebra with unit under the norm

$$\|f\|_1 := \|f\| + \|f'\|, f \in C^1(J).$$

We denote by $L(J)$ the Banach space of all complex-valued Lipschitz continuous functions on J with the following norm

$$\|f\|_L := \sup_{x \in J} |f(x)| + s(f), \tag{1.2.4}$$

with

$$s(f) := \sup_{x \neq y} \frac{|f(x) - f(y)|}{|x - y|} < \infty, \quad f \in L(J).$$

Clearly, $C^1(J) \subset L(J) \subset C(J) \subset B(J)$.

The variation $\operatorname{var}_A f$ over $A \subset J$ of a function $f : J \to \mathbb{C}$ is defined as

$$\sup \sum_{i=1}^{k-1} |f(t_i) - f(t_{i-1})|,$$

the supremum being taken over $t_1 < \cdots < t_k$, $t_i \in A$, $1 \leq i \leq k$, and $k \geq 2$. We write simply $\operatorname{var} f$ for $\operatorname{var}_J f$. If $\operatorname{var} f < \infty$, then f is called a function of bounded variation.

The collection $BV(J)$ of all functions $f : J \to \mathbb{C}$ of bounded variation is a commutative Banach algebra with unit under the norm

$$\|f\|_v := \|f\| + \operatorname{var} f, \, f \in BV(J).$$

Clearly, $L(J) \subset BV(J) \subset B(J)$.

1.3 Markov Chains with an Arbitrary State Space

Let $(X, \mathcal{X})$ and $(Y, \mathcal{Y})$ be two measurable spaces. A function $f : X \to Y$ is said to be $(\mathcal{X}, \mathcal{Y})$-*measurable* if and only if $f^{-1}(A) \in \mathcal{X}$ for every set $A \in \mathcal{Y}$.

Let $(X, \mathcal{X})$ be a measurable space and P *a transition probability function on* $(X, \mathcal{X})$, i.e., $P : X \times \mathcal{X} \to [0, 1]$ is a function such that $P(x, \cdot)$ is a probability measure for all $x \in X$ and $P(\cdot, A)$ is an $(\mathcal{X}, \mathcal{B}_I)$-measurable function for all $A \in \mathcal{X}$.

By Ionescu-Tulcea's theorem [28], for an arbitrarily fixed $x_0 \in X$, there exists a probability space $\left(X^{\mathbb{N}_+}, \mathcal{X}^{\mathbb{N}_+}, P_{x_0}\right)$, where $X^{\mathbb{N}_+}$ is the set of all sequences $(x_1, x_2, \ldots)$ of elements from X, $\mathcal{X}^{\mathbb{N}_+}$ is the smallest σ-algebra containing all the rectangles $A_1 \times \ldots \times A_n \times X \times \ldots$, $A_i \in \mathcal{X}$, $1 \leq i \leq n$, $n \in \mathbb{N}_+$, and P_{x_0} is a probability measure on $\left(X^{\mathbb{N}_+}, \mathcal{X}^{\mathbb{N}_+}\right)$ such that

$$P_{x_0}\left(\operatorname{pr}_{(1,\ldots,n)}^{-1}(A_1 \times \ldots \times A_n)\right)$$
$$= \begin{cases} P(x_0, A_1), & n = 1 \\ \int_{A_1} P(x_0, dx_1) \int_{A_2} P(x_1, dx_2) \ldots \int_{A_n} P(x_{n-1}, dx_n), & n > 1, \end{cases} \quad (1.3.1)$$

for all $A_i \in \mathcal{X}$, $1 \leq i \leq n$, where $\operatorname{pr}_{(1,\ldots,n)} : X^{\mathbb{N}_+} \to X^n$ is *the projection map on the first n coordinates* of $X^{\mathbb{N}_+}$. Define X-valued random variables ξ_n, $n \in \mathbb{N}_+$, on $X^{\mathbb{N}_+}$ by $\xi_n(x) = x_n$, $x = (x_1, x_2, \ldots)$. It follows that

$$P_{x_0}(\xi_1 \in A) = P(x_0, A), \quad A \in \mathcal{X} \quad (1.3.2)$$

$$P_{x_0}(\xi_{n+1} \in A | \xi_1, \ldots, \xi_n) = P(\xi_n, A) \quad P_{x_0}\text{-a.s.} \quad (1.3.3)$$

for all $n \in \mathbb{N}_+$ and $A \in \mathcal{X}$.

The sequence $(\xi_n)_{n \geq 0}$ with $\xi_0 = x_0$ is called a X-*valued homogeneous Markov chain* (or said to have *the Markov property* 1.3.3), whose initial distribution is concentrated at x_0 and whose probability transition function is P.

If X is a finite or countable state space, the transition probability function P is defined as a stochastic matrix $P = (p_{ij})$, $i, j \in X$, with

$$P(i, A) = \sum_{j \in A} p_{ij}, \quad i \in X, A \subset X.$$

1.4 Random Systems with Complete Connections

In this section, we introduce random systems with complete connections and show its properties.

The first explicit formal definition of the concept of dependence with complete connections was given by Onicescu and Mihoc [51]. It is a nontrivial extension of Markovian dependence theory, and it was also investigated by Doeblin and Fortet [19] and by Harris [25]. The concept of random system with complete connections (=RSCC) was defined by Iosifescu [26].

Examples of RSCC are urns models [32, 51], stochastic learning processes [32, 33, 50], partially observed random chains [32], image coding [5], continued fraction expansions [28, 38, 40, 42, 59, 64], and others.

An RSCC is often called an *iterated function system with place-dependent probabilities*. In the last three decades a lot of work has been devoted to *iterated function systems* (=IFS). IFS is not a new concept. It only became fashionable in the framework of fractals [4] and chaos [3], but, before that, it appeared as the simplest case of an RSCC.

For applications of RSCC to continued fraction expansions, see also [31, 59, 60, 64].

Definition 1.4.1 ([28]) A random system with complete connections is a quadruple

$$\{(W, \mathcal{W}), (X, \mathcal{X}), u, P\}, \tag{1.4.1}$$

where:

(i) $(W, \mathcal{W})$ and $(X, \mathcal{X})$ are arbitrary measurable spaces.
(ii) $u : W \times X \to W$ is a $(\mathcal{W} \otimes \mathcal{X}, \mathcal{W})$ measurable map.
(iii) P is a transition probability function from $(W, \mathcal{W})$ to $(X, \mathcal{X})$, i.e., $P : W \times \mathcal{X} \to [0, 1]$ is a function such that $P(w, \cdot)$ is a probability measure for all $w \in W$, and $P(\cdot, A)$ is a measurable function on $(W, \mathcal{W})$ for all $A \in \mathcal{X}$.

For an RSCC in the above definition we call W the *state space*, X the *event space*, and u the *transition function*. The role of the function u is to change a state $w \in W$ into the new state $w' = u(w, x) \in W$ by an event $x \in X$

$$W \ni w \overset{x}{\mapsto} w' = u(w, x) \in W.$$

In this case, $P(w, x)$ is regarded as the probability of the transition $w \mapsto w'$ which depends on the information of both w and x.

For any $n \in \mathbb{N}_+$, consider the maps $u^{(n)} : W \times X^n \to W$, defined by

$$u^{(1)}(w, x) := u(w, x),$$
$$u^{(n+1)}\left(w, x^{(n+1)}\right) := u\left(u^{(n)}\left(w, x^{(n)}\right), x_{n+1}\right), \ n \geq 1,$$

where $x^{(n)} = (x_1, \ldots, x_n) \in X^n$. We will simply write $wx^{(n)}$ for $u^{(n)}(w, x^{(n)})$. For every $w \in W, r \in \mathbb{N}_+$ and $A \in \mathcal{X}^r$, define

$$P_1(w, A) := P(w, A),$$
$$P_r(w, A) := \int_X P(w, dx_1) \int_X P(wx_1, dx_2) \ldots \int_X P(wx^{(r-1)}, dx_r)\chi_A(x^{(r)}), \ r \geq 2,$$
(1.4.2)

where χ_A is the characteristic function of the set A. Obviously, for $r \in \mathbb{N}_+$ fixed, P_r is a transition probability function from $(W, \mathcal{W})$ to $(X^r, \mathcal{X}^r)$.

By virtue of the existence theorem ([28], Theorem 1.1.2), for a given RSCC $\{(W, \mathcal{W}), (X, \mathcal{X}), u, P\}$ and for an arbitrarily fixed element $w \in W$, one can generate two stochastic sequences $(\xi_n)_{n \geq 0}$ and $(\zeta_n)_{n \geq 1}$ as follows: We set $\xi_0 = w$, pick an element $\zeta_1 \in X$ using $P(\xi_0, \cdot)$, define $\xi_1 = u(\xi_0, \zeta_1)$, pick ζ_2 in X using $P(\xi_1, \cdot)$, define $\xi_2 = u(\xi_1, \zeta_2)$, and generally we pick ζ_n in X using $P(\xi_{n-1}, \cdot)$, and define $\xi_n = u(\xi_{n-1}, \zeta_n)$. So, in fact, we have

$$\xi_0 = w, \ \xi_{n+1} = u(\xi_n, \zeta_{n+1}), \ n \geq 1,$$
$$P(\zeta_1 \in A) = P(w, A), \ A \in \mathcal{X}$$
$$P(\zeta_{n+1} \in A | \xi_n, \zeta_n, \ldots, \xi_1, \zeta_1, \xi_0) = P(\xi_n, A), \ A \in \mathcal{X}.$$

We call the sequence $(\xi_n)_{n \geq 0}$ of W-valued random variables the *state sequence* and the sequence $(\zeta_n)_{n \geq 1}$ of X-valued random variables the *event sequence*.

Obviously, $(\zeta_n)_{n \geq 1}$ is no longer Markovian, but a *chain with complete connections* (whose transition probabilities depend on the whole past history).

The sequence $(\xi_n)_{n \geq 0}$ is a Markov chain (the so-called *associated Markov chain*) with the transition operator U defined by

$$Uf(w) := \int_X P(w, dx) f(wx), \quad f \in B(W, \mathcal{W}),$$

where $B(W, \mathcal{W})$ is the Banach space of all bounded $\mathcal{W}$-measurable complex-valued functions defined on W. Moreover, the transition probability function of the associated

1.4 Random Systems with Complete Connections

Markov chain is

$$Q(w, B) := \int_X P(w, dx)\chi_B(wx) = P(w, B_w),$$

where $B_w = \{x \in X : wx \in B\}$, $w \in W$, $B \in \mathcal{W}$. The iterates of the operator U are given by

$$U^n f(w) = \int_{X^n} P_n(w, dx^{(n)}) f(wx^{(n)}), \quad f \in B(W, \mathcal{W}), n \in \mathbb{N}_+.$$

It follows that the n-step transition probability function is given by

$$Q^n(w, B) = P_n(w, B_w^{(n)}), \quad w \in W, \ B \in \mathcal{W}, n \in \mathbb{N}_+, \quad (1.4.3)$$

where $B_w^{(n)} = \{x^{(n)} : wx^{(n)} \in B\}$. Hence the transition operator associated with the Markov chain with state space $(W, \mathcal{W})$ and transition probability function Q is defined by

$$Uf(\cdot) := \int_W Q(\cdot, dw) f(w), \quad f \in B(W, \mathcal{W}). \quad (1.4.4)$$

Its iterates are given by

$$U^n f(\cdot) = \int_W Q^n(\cdot, dw) f(w), \quad n \in \mathbb{N}_+, \quad (1.4.5)$$

where Q^n is the n-step transition probability function.

Now define

$$Q_n(w, B) = \frac{1}{n} \sum_{k=1}^n Q^k(w, B), \quad (w, B) \in W \times \mathcal{W}, \ n \geq 1.$$

So $Q_1 = Q$. Then $(W, \mathcal{W}, Q_n(w, \cdot))$ is also a probability space for each $w \in W$. Let U_n be the Markov operator associated with Q_n, i.e.,

$$U_n f(w) := \int_W Q_n(w, dw') f(w') \quad f \in B(W, \mathcal{W}). \quad (1.4.6)$$

Following [28], we introduce several characterizations of the operator U as follows.

Definition 1.4.2 Let $W, U, U_n, L(W)$ be as in (1.4.1), (1.4.4), (1.4.6), and (1.2.4), respectively. Assuming that (W, d) is a metric space, we say:

(i) U is ordered if there exists a bounded linear operator U^∞ on $L(W)$ such that

$$\lim_{n \to \infty} \|U_n f - U^\infty f\|_L = 0, \quad f \in L(W). \tag{1.4.7}$$

(ii) U is aperiodic if there exists a bounded linear operator U^∞ on $L(W)$ such that

$$\lim_{n \to \infty} \|U^n f - U^\infty f\|_L = 0, \quad f \in L(W), \tag{1.4.8}$$

where U^n is the n-th iterate of U for $n \geq 1$.
(iii) U is ergodic with respect to $L(W)$ if U is ordered and the rank of U^∞ in (1.4.7) is 1.
(iv) U is regular with respect to $L(W)$ if U is ergodic and aperiodic.
(v) The Markov chain corresponding to U is regular if U is regular with respect to $L(W)$.
(vi) The Markov chain corresponding to U is ordered if U is ordered with respect to $L(W)$.

Definition 1.4.3 The transition operator U of a Markov chain with state space (W, d) is said to be a Doeblin-Fortet operator if and only if U maps $L(W)$ into $L(W)$ boundedly with respect to $\|\cdot\|_L$ and there exist $k \in \mathbb{N}_+$, $\alpha \in [0, 1)$, and $\beta < \infty$ such that

$$s(U^k f) \leq \alpha \cdot s(f) + \beta \cdot \|f\|, \quad f \in L(W). \tag{1.4.9}$$

Alternatively, the Markov chain itself is said to be a *Doeblin-Fortet chain*.

Definition 1.4.4 A Markov chain is said to be compact if and only if its state space is a compact metric space (W, d) and its transition operator is a Doeblin-Fortet operator.

Theorem 1.4.5 *The Markov chain associated with an RSCC with contraction is a Doeblin-Fortet chain.*

Theorem 1.4.6

(i) *A compact Markov chain is ordered with respect to $L(W)$, and there exists a transition probability function Q^∞ on $(W, \mathcal{W})$ such that*

$$U^\infty f(\cdot) = \int_W Q^\infty(\cdot, dw) f(w), \quad f \in L(W). \tag{1.4.10}$$

1.4 Random Systems with Complete Connections

Moreover, for any $w \in W$, $Q^\infty(w, \cdot)$ is a stationary probability for the chain, i.e.,

$$\int_W Q^\infty(w, \mathrm{d}w') Q(w', B) = Q^\infty(w, B), \quad w \in W, B \in \mathcal{W}. \tag{1.4.11}$$

(ii) A compact Markov chain is ergodic with respect to $L(W)$ if and only if it possesses a unique stationary probability.

Lemma 1.4.7 *Assume that the Markov operator U is aperiodic with respect to $L(W)$. Then we have $U^\infty U^\infty = U^\infty$, $(U - U^\infty) U^\infty = U^\infty (U - U^\infty) = 0$, and $U^n = U^\infty + (U - U^\infty)^n$ for $n \in \mathbb{N}_+$. Moreover, there exist positive constants $q < 1$ and K such that*

$$\|U^n f - U^\infty f\|_L \leq K q^n \|f\|_L, \quad f \in L(W), n \in \mathbb{N}_+. \tag{1.4.12}$$

Definition 1.4.8 ([50]) The system $\{(W, \mathcal{W}), (X, \mathcal{X}), u, P\}$ is an RSCC with contraction if (W, d) is a separable metric space and the following conditions are satisfied:

(i) $\vartheta_1 < \infty$.
(ii) $\vartheta_\ell < 1$ for some $\ell \geq 1$.
(iii) $\Upsilon_1 < \infty$

with

$$\vartheta_k := \sup_{w' \neq w''} \int_{X^k} P_k\left(w', \mathrm{d}x^{(k)}\right) \frac{d\left(w'x^{(k)}, w''x^{(k)}\right)}{d(w', w'')}, \tag{1.4.13}$$

$$\Upsilon_k := \sup_{A \in \mathcal{X}^k} \sup_{w' \neq w''} \frac{|P_k(w', A) - P_k(w'', A)|}{d(w', w'')}, \quad k \in \mathbb{N}_+, \tag{1.4.14}$$

where P_k is a transition probability function from $(W, \mathcal{W})$ to $(X^k, \mathcal{X}^k)$ defined in (1.4.2).

Theorem 1.4.9 *Let $\{(W, \mathcal{W}), (X, \mathcal{X}), u, P\}$ be an RSCC with contraction. For the n-step transition probability function Q^n, define $\{\sigma_n\}$ by*

$$\sigma_n(w) := \operatorname{supp} Q^n(w, \cdot), \quad w \in W, \tag{1.4.15}$$

where $\operatorname{supp} \mu$ denotes the support of a measure μ. Assume that (W, d) is a compact metric space. Then the following holds:

(i) The Markov chain associated with the RSCC is regular if and only if there exists a point $w_0 \in W$ such that

$$\lim_{n \to \infty} d(\sigma_n(w), w_0) = 0 \quad \text{for all } w \in W, \tag{1.4.16}$$

where $d(A, w) := \inf_{w' \in A} d(w', w)$ for $A \subset W$.

(ii) For $\{\sigma_n\}$ in (1.4.15), we have

$$\sigma_{m+n}(w) = \overline{\bigcup_{w' \in \sigma_m(w)} \sigma_n(w')}, \tag{1.4.17}$$

for all $m, n \in \mathbb{N}_+$, $w \in W$. Here the overline means the topological closure in W.

2 θ-Continued Fraction Expansions

S. Chakraborty and B.V. Rao [10] have introduced the continued fraction expansion of a number in terms of an irrational $\theta \in (0, 1)$. This new expansion of positive reals, different from the regular continued fraction expansion, is called θ-*continued fraction expansion* or θ-*expansion* for short. We mention that the case $\theta = 1$ refers to regular continued fraction (RCF) expansions. The study initiated by S. Chakraborty and B.V. Rao on the analogous transformation of the Gauss map was completed by P.S. Chakraborty and B.V. Dasgupta. Actually, in [9] was identified the absolutely continuous invariant probability measure of this new transformation only in the particular case $\theta^2 = 1/m$, $m \in \mathbb{N}_+$.

2.1 θ-Continued Fraction Expansion as Dynamical System

For a fixed $\theta \in (0, 1)$, S. Chakraborty and B.V. Rao [10] showed that any $x \in (0, \theta)$ can be written in the form

$$x = \cfrac{1}{b_1\theta + \cfrac{1}{b_2\theta + \cfrac{1}{b_3\theta + \cdots}}} =: [b_1\theta, b_2\theta, b_3\theta, \ldots] \qquad (2.1.1)$$

which is called the θ-*continued fraction expansion* of x. Here b_n's are nonnegative integers and are called θ-*expansion digits*.

Here and hereafter, we ignore the index θ in the θ-expansion digit notation, and we simply write b_n instead of $b_{\theta,n}$.

This continued fraction is treated as the following dynamical system.

© The Author(s), under exclusive license to Springer Nature Switzerland AG 2025
G. I. Sebe, D. Lascu, *Metrical and Ergodic Theory of Continued Fraction Algorithms*,
Frontiers in Mathematics, https://doi.org/10.1007/978-3-031-86634-0_2

Definition 2.1.1 Fix an irrational $\theta \in (0, 1)$ and $m \in \mathbb{N}_+$ such that $\theta^2 = 1/m$:

(i) The measure-theoretical dynamical system $([0, \theta], \mathcal{B}_{[0,\theta]}, T_\theta)$ is defined as follows: $\mathcal{B}_{[0,\theta]}$ denotes the σ-algebra of all Borel subsets of $[0, \theta]$, and T_θ is the transformation

$$T_\theta : [0, \theta] \to [0, \theta]; \quad T_\theta(x) := \begin{cases} \dfrac{1}{x} - \theta \left\lfloor \dfrac{1}{\theta x} \right\rfloor & \text{if } x \in (0, \theta], \\ 0 & \text{if } x = 0. \end{cases} \quad (2.1.2)$$

(ii) In addition to (i), we write $([0, \theta], \mathcal{B}_{[0,\theta]}, \gamma_\theta, T_\theta)$ as $([0, \theta], \mathcal{B}_{[0,\theta]}, T_\theta)$ with the following probability measure γ_θ on $([0, \theta], \mathcal{B}_{[0,\theta]})$:

$$\gamma_\theta(A) := \frac{1}{\log(1 + \theta^2)} \int_A \frac{\theta \, dx}{1 + \theta x}, \quad A \in \mathcal{B}_{[0,\theta]}. \quad (2.1.3)$$

By using T_θ, the sequence $(b_n)_{n \in \mathbb{N}_+}$ in (2.1.1) is obtained as follows:

$$b_n = b_n(x) := b_1\left(T_\theta^{n-1}(x)\right), \quad n \geq 1, \quad (2.1.4)$$

with $T_\theta^0(x) = x$ and

$$b_1 = b_1(x) := \begin{cases} \left\lfloor \dfrac{1}{\theta x} \right\rfloor & \text{if } x \neq 0, \\ \infty & \text{if } x = 0. \end{cases} \quad (2.1.5)$$

In this way, T_θ algorithmically generates the θ-expansion.

Proposition 2.1.2 Let $([0, \theta], \mathcal{B}_{[0,\theta]}, \gamma_\theta, T_\theta)$ be as in Definition 2.1.1(ii):

(i) $([0, \theta], \mathcal{B}_{[0,\theta]}, \gamma_\theta, T_\theta)$ is ergodic.
(ii) The measure γ_θ is invariant under T_θ, that is, $\gamma_\theta(A) = \gamma_\theta(T_\theta^{-1}(A))$ for any $A \in \mathcal{B}_{[0,\theta]}$.

Proof See Section 8 in [10]. □

2.2 Basic Metric Properties

Define the n-th order convergent $[b_1\theta, b_2\theta, \ldots, b_n\theta]$ of $x \in [0, \theta]$ by truncating the θ-expansion in (2.1.1). Thus, S. Chakraborty and B.V. Rao proved in [10] that

$$[b_1\theta, b_2\theta, \ldots, b_n\theta] \to x, \quad n \to \infty. \quad (2.2.1)$$

2.2 Basic Metric Properties

In what follows the stated identities hold for all n in case x has an infinite θ-expansion and they hold for $n \leq k$ in case x has a finite θ-expansion terminating at the k-th stage.

Here and below we omit the index θ whenever it is clear from context.

Define real functions $p_{\theta,n}(x) = p_n(x)$ and $q_{\theta,n}(x) = q_n(x)$, for $n \in \mathbb{N}_+$, by

$$p_n(x) := b_n(x)\theta p_{n-1}(x) + p_{n-2}(x), \qquad (2.2.2)$$

$$q_n(x) := b_n(x)\theta q_{n-1}(x) + q_{n-2}(x), \qquad (2.2.3)$$

with $p_{-1}(x) = 1$, $p_0(x) = 0$, $q_{-1}(x) = 0$, and $q_0(x) = 1$. By induction, we have

$$p_{n-1}(x)q_n(x) - p_n(x)q_{n-1}(x) = (-1)^n, \quad n \in \mathbb{N}. \qquad (2.2.4)$$

By using (2.2.2) and (2.2.3), we can verify that

$$x = \frac{p_n(x) + T_\theta^n(x) p_{n-1}(x)}{q_n(x) + T_\theta^n(x) q_{n-1}(x)}, \quad n \geq 1. \qquad (2.2.5)$$

By taking $T_\theta^n(x) = 0$ in (2.2.5), we obtain $[b_1\theta, b_2\theta, \ldots, b_n\theta] = p_n(x)/q_n(x)$. Using (2.2.4) and (2.2.5) we obtain

$$x - \frac{p_n(x)}{q_n(x)} = \frac{(-1)^{n+1} T_\theta^n(x)}{q_n(x)(q_n(x) + T_\theta^n(x) q_{n-1}(x))}, \quad n \geq 1. \qquad (2.2.6)$$

By applying $0 \leq T_\theta^n \leq \theta$ to (2.2.6), we can verify that

$$\frac{1}{q_n(x)(q_{n+1}(x) + \theta q_n(x))} \leq \left| x - \frac{p_n(x)}{q_n(x)} \right| \leq \frac{1}{q_n(x) q_{n+1}(x)}. \qquad (2.2.7)$$

From (2.2.3), we have that $q_n(x) \geq \theta$, $n \in \mathbb{N}_+$. Further, also from (2.2.3) and by induction we have that

$$q_n(x) \geq \left\lfloor \frac{n}{2} \right\rfloor \theta^2. \qquad (2.2.8)$$

Finally, (2.2.1) follows from (2.2.7) and (2.2.8).

Putting $\mathbb{N}_m := \{m, m+1, \ldots\}$, $m \in \mathbb{N}_+$, the incomplete quotients b_n, $n \in \mathbb{N}_+$, take positive integer values in $\mathbb{N}_m$. We now introduce a partition of the interval $[0, \theta]$ which is natural to the θ-expansions. Such a partition is generated by the *fundamental intervals* (or *cylinders*) of rank n. For any $n \in \mathbb{N}_+$ and $i^{(n)} = (i_1, \ldots, i_n) \in \mathbb{N}_m^n$, define the *fundamental interval associated with* $i^{(n)}$ by

$$I_\theta\left(i^{(n)}\right) = \{x \in [0, \theta] : b_k(x) = i_k \text{ for } k = 1, \ldots, n\}, \qquad (2.2.9)$$

where $I_\theta \left(i^{(0)}\right) = [0, \theta]$. For example, for any $i \in \mathbb{N}_m$ we have

$$I_\theta(i) = \{x \in [0, \theta] : b_1(x) = i\} = \left(\frac{1}{(i+1)\theta}, \frac{1}{i\theta}\right]. \quad (2.2.10)$$

From the definition of T_θ and (2.2.5), we have

$$I_\theta\left(b^{(n)}\right) = \left(u_\theta\left(b^{(n)}\right), v_\theta\left(b^{(n)}\right)\right), \quad (2.2.11)$$

where $u_\theta\left(b^{(n)}\right)$ and $v_\theta\left(b^{(n)}\right)$ are defined as

$$u_\theta\left(b^{(n)}\right) := \begin{cases} \dfrac{p_n + \theta p_{n-1}}{q_n + \theta q_{n-1}} & \text{if } n \text{ is odd}, \\ \dfrac{p_n}{q_n} & \text{if } n \text{ is even}, \end{cases} \quad (2.2.12)$$

and

$$v_\theta\left(b^{(n)}\right) := \begin{cases} \dfrac{p_n}{q_n} & \text{if } n \text{ is odd}, \\ \dfrac{p_n + \theta p_{n-1}}{q_n + \theta q_{n-1}} & \text{if } n \text{ is even}, \end{cases} \quad (2.2.13)$$

where $p_n := p_n(x)$ and $q_n := q_n(x)$ are defined in (2.2.2) and (2.2.3), respectively.

Let λ_θ denote the Lebesgue measure on $[0, \theta]$. Using (2.2.4) we get

$$\lambda_\theta\left(I_\theta\left(b^{(n)}\right)\right) = \frac{1}{\theta}\left|\frac{p_n}{q_n} - \frac{p_n + \theta p_{n-1}}{q_n + \theta q_{n-1}}\right| = \frac{1}{q_n(q_n + \theta q_{n-1})}. \quad (2.2.14)$$

To derive the so-called Brodén-Borel-Lévy formula [28, 29] for θ-expansions, let us define $(s_{\theta,n})_{n \in \mathbb{N}}$ by

$$s_{\theta,0} := 0, \quad s_{\theta,n} := \frac{q_{n-1}}{q_n}, \quad n \geq 1. \quad (2.2.15)$$

From (2.2.3), $s_{\theta,n} = 1/(b_n\theta + s_{\theta,n-1})$ for $n \geq 1$. Hence

$$s_{\theta,n} = \cfrac{1}{b_n\theta + \cfrac{1}{b_{n-1}\theta + \cdots + \cfrac{1}{b_1\theta}}} = [b_n\theta, b_{n-1}\theta, \ldots, b_1\theta], \quad (2.2.16)$$

for $n \geq 1$.

2.2 Basic Metric Properties

Proposition 2.2.1 (Brodén-Borel-Lévy-Type Formula [64]) *Let λ_θ denote the Lebesgue measure on $[0, \theta]$. For any $n \in \mathbb{N}_+$, the conditional probability $\lambda_\theta(T_\theta^n < x | b_1, \ldots, b_n)$ is given as follows:*

$$\lambda_\theta(T_\theta^n < x | b_1, \ldots, b_n) = \frac{(s_{\theta,n}\theta + 1)x}{\theta(s_{\theta,n}x + 1)}, \quad x \in [0, \theta], \tag{2.2.17}$$

where $s_{\theta,n}$ is defined in (2.2.15) and b_n's are as in (2.1.4).

Proof By definition, we have

$$\lambda_\theta(T_\theta^n < x | b_1, \ldots, b_n) = \frac{\lambda_\theta\left((T_\theta^n < x) \cap I_\theta(b_1, \ldots, b_n)\right)}{\lambda_\theta\left(I_\theta(b_1, \ldots, b_n)\right)} \tag{2.2.18}$$

for any $n \in \mathbb{N}_+$ and $x \in [0, \theta]$. Using (2.2.5) and (2.2.11) we get

$$\lambda_\theta\left((T_\theta^n < x) \cap I_\theta(b_1, \ldots, b_n)\right) = \frac{1}{\theta}\left|\frac{p_n}{q_n} - \frac{p_n + xp_{n-1}}{q_n + xq_{n-1}}\right| = \frac{x}{q_n(q_n + xq_{n-1})\theta}.$$

From this and (2.2.14) it follows that

$$\lambda_\theta\left(T_\theta^n < x | b_1, \ldots, b_n\right) = \frac{\lambda_\theta\left((T_\theta^n < x) \cap I_\theta(b_1, \ldots, b_n)\right)}{\lambda_\theta\left(I_\theta(b_1, \ldots, b_n)\right)}$$

$$= \frac{x(q_n + \theta q_{n-1})}{(q_n + xq_{n-1})\theta} = \frac{x(s_{\theta,n}\theta + 1)}{(s_{\theta,n}x + 1)\theta},$$

for any $n \in \mathbb{N}_+$ and $x \in [0, \theta]$. □

The Brodén-Borel-Lévy formula allows us to determine the probability structure of incomplete quotients $(b_n)_{n \in \mathbb{N}_+}$ under λ_θ.

Proposition 2.2.2 ([64]) *For any $i \in \mathbb{N}_m$ and $n \in \mathbb{N}_+$, we have*

$$\lambda_\theta(b_1 = i) = \frac{m}{i(i+1)}, \quad \lambda_\theta(b_{n+1} = i | b_1, \ldots, b_n) = P_{\theta,i}(s_{\theta,n}), \tag{2.2.19}$$

where $(s_{\theta,n})_{n \in \mathbb{N}_+}$ is defined in (2.2.15), and

$$P_{\theta,i}(x) := \frac{x\theta + 1}{(x + i\theta)(x + (i+1)\theta)}. \tag{2.2.20}$$

Proof From (2.2.10), the case $\lambda_\theta(b_1 = i)$ holds. Using (2.1.4) and (2.2.17) we obtain

$$\lambda_\theta(b_{n+1} = i \mid b_1, \ldots, b_n) = \lambda_\theta \left(T_\theta^n \in \left(\frac{1}{(i+1)\theta}, \frac{1}{i\theta} \right] \mid b_1, \ldots, b_n \right).$$

$$= \frac{(s_{\theta,n}\theta + 1)\frac{1}{i\theta}}{\theta(s_{\theta,n}\frac{1}{i\theta} + 1)} - \frac{(s_{\theta,n}\theta + 1)\frac{1}{(i+1)\theta}}{\theta(s_{\theta,n}\frac{1}{(i+1)\theta} + 1)} = P_{\theta,i}(s_{\theta,n}).$$

$\square$

Remark 2.2.3

(i) It is easy to check that $\sum_{i \geq m} P_{\theta,i}(x) = 1$ for any $x \in [0, \theta]$.
(ii) Proposition 2.2.2 is the starting point of an approach to the metrical theory of θ-expansions via dependence with complete connections (see [28], Section 5.2).

Corollary 2.2.4 ([64]) *The sequence* $(s_{\theta,n})_{n \in \mathbb{N}_+}$ *with* $s_{\theta,0} = 0$ *is an* $[0, \theta]$-*Markov chain on* $([0, \theta], \mathcal{B}_{[0,\theta]}, \lambda_\theta)$ *with the following transition mechanism: From state s the possible transitions are to any state* $1/(s + i\theta)$ *with corresponding transition probability* $P_{\theta,i}(s)$, $i \in \mathbb{N}_m$.

2.3 Some Ergodic Properties

In this section, we give some ergodic properties of θ-expansions, and we find the entropy of the transformation which generates θ-expansion.

Using the ergodicity of T_θ and Birkhoff's ergodic theorem [14], a number of results were obtained in [10].

For q_n in (2.2.3), its asymptotic growth rate β_θ is defined as

$$\beta_\theta := \lim_{n \to \infty} \frac{1}{n} \log q_n = \frac{-1}{\log(1+\theta^2)} \int_0^\theta \frac{\theta \log x}{1 + \theta x} dx. \tag{2.3.1}$$

This is a Lévy result. We also have a Khintchine result, i.e., the asymptotic value of the arithmetic mean of $b_1, b_2, \ldots, b_n$, where b_n's are given in (2.1.4). We have

$$\lim_{n \to \infty} \frac{b_1 + b_2 + \ldots + b_n}{n} = \infty. \tag{2.3.2}$$

Let $\theta \in (0, 1)$ and $m \in \mathbb{N}_+$ such that $\theta^2 = 1/m$. In this section we give some consequences of ergodicity in terms of properties of the continued fraction expansion for almost every real number $x \in (0, \theta)$.

2.3 Some Ergodic Properties

The following theorem presents other Lévy and Khintchine-type results. The relation (2.3.1) allows to find other asymptotical results of this type. Applying our machinery to the ergodic system $([0, \theta], \mathcal{B}_{[0,\theta]}, \gamma_\theta, T_\theta)$, we obtain a result on the statistical data of θ-expansions, i.e., almost sure asymptotics for the geometrical mean value for partial quotients.

Proposition 2.3.1 *For almost all $x \in (0, \theta)$ one has*

$$\lim_{n \to \infty} \frac{1}{n} \log \left(\lambda_\theta \left(I_\theta \left(i^{(n)} \right) \right) \right) = -2\beta_\theta, \tag{2.3.3}$$

$$\lim_{n \to \infty} \frac{1}{n} \log \left| x - \frac{p_n}{q_n} \right| = -2\beta_\theta, \tag{2.3.4}$$

$$\lim_{n \to \infty} (b_1 b_2 \cdots b_n)^{1/n} = \prod_{k \geq m} \left(1 + \frac{1}{k(k+2)} \right)^{\frac{\log k}{\log(1+\theta^2)}}. \tag{2.3.5}$$

Here, λ_θ denotes the Lebesgue measure on $[0, \theta]$ and $I_\theta \left(i^{(n)} \right)$ are the cylinders in (2.2.9), and b_n's, p_n, and q_n are given in (2.1.4),(2.2.2), and (2.2.3), respectively.

Proof

(i) From (2.2.14) we have

$$\frac{1}{(1+\theta)q_n^2} < \lambda_\theta \left(I_\theta \left(i^{(n)} \right) \right) < \frac{1}{q_n^2}, \tag{2.3.6}$$

or

$$-\log(1+\theta) - 2\log(q_n) < \log \lambda_\theta \left(I_\theta \left(i^{(n)} \right) \right) < -2\log(q_n). \tag{2.3.7}$$

Now apply (2.3.1) to obtain (2.3.3).

(ii) (2.3.4) follows from (2.3.1) and 2.2.7.

(iii) Let $f(x) := \log(b_1(x))$ for $x \in (0, \theta)$, where b_1 is as in (2.1.5). Then $f \in L^1((0,\theta), \gamma_\theta)$, i.e., f is an integrable function, since

$$\int_0^\theta f \, d\gamma_\theta = \sum_{k \geq m} \int_{\frac{1}{(k+1)\theta}}^{\frac{1}{k\theta}} f \, d\gamma_\theta = \sum_{k \geq m} \frac{1}{\log(1+\theta^2)} \int_{\frac{1}{(k+1)\theta}}^{\frac{1}{k\theta}} \frac{\theta \log b_1(x)}{1+\theta x} dx$$

$$= \frac{1}{\log(1+\theta^2)} \sum_{k \geq m} \log k \cdot \log \left(1 + \frac{1}{k(k+2)} \right)$$

$$= \frac{1}{\log(1+\theta^2)} \sum_{k \geq m} \frac{\log k}{k(k+2)}. \tag{2.3.8}$$

Since the series $\sum_{k \geq m} (\log k)/(k(k+2))$ is convergent, it follows that $\int_0^\theta f \, d\gamma_\theta := s \in \mathbb{R}$. Now we have

$$\lim_{n \to \infty} (b_1 b_2 \cdots b_n)^{1/n} = e^s = \prod_{k \geq m} \left(1 + \frac{1}{k(k+2)}\right)^{\frac{\log k}{\log(1+\theta^2)}}. \tag{2.3.9}$$

$\square$

2.4 An Infinite-Order-Chain Representation

In this section we introduce the natural extension $\overline{T}_\theta$ of T_θ in (2.1.2) and define extended random variables according to [29, Chapter 1.3.3]. Then we give an infinite-order-chain representation of the sequence of the incomplete quotients for θ-expansions.

2.4.1 Natural Extension of T_θ

Let $\left([0, \theta], \mathcal{B}_{[0,\theta]}, T_\theta\right)$ be as above. Define $(u_{\theta,i})_{i \in \mathbb{N}_m}$ by

$$u_{\theta,i} : [0, \theta] \to [0, \theta]; \quad u_{\theta,i}(x) := \frac{1}{x + i\theta}, \quad x \in [0, \theta]. \tag{2.4.1}$$

For each $i \in \mathbb{N}_m$, $u_{\theta,i}$ is a right inverse of T_θ, that is,

$$(T_\theta \circ u_{\theta,i})(x) = x, \quad \text{for any } x \in [0, \theta]. \tag{2.4.2}$$

Furthermore, if $b_1(x) = i$, then $(u_{\theta,i} \circ T_\theta)(x) = x$ where b_1 is as in (2.1.5).

Definition 2.4.1 The *natural extension* $\left([0, \theta]^2, \mathcal{B}^2_{[0,\theta]}, \overline{T}_\theta\right)$ of $\left([0, \theta], \mathcal{B}_{[0,\theta]}, T_\theta\right)$ is the transformation $\overline{T}_\theta$ of the square space $\left([0, \theta]^2, \mathcal{B}^2_{[0,\theta]}\right) := \left([0, \theta], \mathcal{B}_{[0,\theta]}\right) \times \left([0, \theta], \mathcal{B}_{[0,\theta]}\right)$ defined as follows [49]:

$$\overline{T}_\theta : [0, \theta]^2 \to [0, \theta]^2; \quad \overline{T}_\theta(x, y) := \left(T_\theta(x), u_{\theta, b_1(x)}(y)\right), \quad (x, y) \in [0, \theta]^2. \tag{2.4.3}$$

2.4 An Infinite-Order-Chain Representation

From (2.4.2), we see that $\overline{T}_\theta$ is bijective on $[0, \theta]^2$ with the inverse

$$(\overline{T}_\theta)^{-1}(x, y) = \left(u_{\theta, b_1(y)}(x), T_\theta(y)\right), \quad (x, y) \in [0, \theta]^2. \tag{2.4.4}$$

Iterations of (2.4.3) and (2.4.4) are given as follows for each $n \geq 2$:

$$\left(\overline{T}_\theta\right)^n (x, y) = \left(T_\theta^n(x), [b_n(x)\theta, b_{n-1}(x)\theta, \ldots, b_2(x)\theta, b_1(x)\theta + y]\right), \tag{2.4.5}$$

$$\left(\overline{T}_\theta\right)^{-n} (x, y) = \left([b_n(y)\theta, b_{n-1}(y)\theta, \ldots, b_2(y)\theta, b_1(y)\theta + x], T_\theta^n(y)\right). \tag{2.4.6}$$

For γ_θ in (2.1.3), define its *extended measure* $\overline{\gamma}_\theta$ on $\left([0, \theta]^2, \mathcal{B}_{[0,\theta]}^2\right)$ as

$$\overline{\gamma}_\theta(B) := \frac{1}{\log(1 + \theta^2)} \iint_B \frac{dxdy}{(1 + xy)^2}, \quad B \in \mathcal{B}_{[0,\theta]}^2. \tag{2.4.7}$$

Then $\overline{\gamma}_\theta(A \times [0, \theta]) = \overline{\gamma}_\theta([0, \theta] \times A) = \gamma_\theta(A)$ for any $A \in \mathcal{B}_{[0,\theta]}$.

The measure $\overline{\gamma}_\theta$ is preserved by $\overline{T}_\theta$ [64], i.e., $\overline{\gamma}_\theta((\overline{T}_\theta)^{-1}(B)) = \overline{\gamma}_\theta(B)$ for any $B \in \mathcal{B}_{[0,\theta]}^2$. Since $\overline{T}_\theta$ is invertible on $[0, \theta]^2$, the last equation is equivalent to

$$\overline{\gamma}_\theta\left(\overline{T}_\theta(B)\right) = \overline{\gamma}_\theta(B), \quad \text{for any } B \in \mathcal{B}_{[0,\theta]}^2. \tag{2.4.8}$$

2.4.2 Extended Random Variables

With respect to $\overline{T}_\theta$ in (2.4.3), define *extended incomplete quotients* $\overline{b}_l(x, y), l \in \mathbb{Z} := \{\ldots, -2, -1, 0, 1, 2, \ldots\}$ at $(x, y) \in [0, \theta]^2$ by

$$\overline{b}_{l+1}(x, y) := \overline{b}_1\left((\overline{T}_\theta)^l(x, y)\right), \quad l \in \mathbb{Z}, \tag{2.4.9}$$

with $\overline{b}_1(x, y) = b_1(x)$, $x, y \in [0, \theta]$. Since $\overline{T}_\theta$ is invertible, it follows that $\overline{b}_l(x, y)$ in (2.4.9) is also well-defined for $l \leq 0$. By (2.3.3), we have

$$\overline{b}_n(x, y) = b_n(x), \quad \overline{b}_0(x, y) = b_1(y), \quad \overline{b}_{-n}(x, y) = b_{n+1}(y), \tag{2.4.10}$$

for any $n \in \mathbb{N}_+$ and $(x, y) \in [0, \theta]^2$. Since $\overline{\gamma}_\theta$ is preserved by $\overline{T}_\theta$, the doubly infinite sequence $(\overline{b}_l(x, y))_{l \in \mathbb{Z}}$ is strictly stationary (i.e., its distribution is invariant under a shift of the indices) under $\overline{\gamma}_\theta$. The stochastic property of $(\overline{b}_l)_{l \in \mathbb{Z}}$ follows from the fact that

$$\overline{\gamma}_\theta([0, x] \times [0, \theta] \mid \overline{b}_0, \overline{b}_{-1}, \ldots) = \frac{(b\theta + 1)x}{(bx + 1)\theta} \quad \overline{\gamma}_\theta\text{-a.s.}, \tag{2.4.11}$$

for any $x \in [0, \theta]$, where $\overline{b} := [\overline{b}_0 \theta, \overline{b}_{-1} \theta, \ldots]$ with $\overline{b}_l := \overline{b}_l(x, y)$ for $l \in \mathbb{Z}$ and $(x, y) \in [0, \theta]^2$. If $I_{\theta,n}$ denotes the n-th order cylinder $I_\theta(b_1, b_2, \ldots, b_n)$, since $I_\theta(\overline{b}_0, \overline{b}_{-1}, \ldots, \overline{b}_{-n}) = [0, \theta] \times I_{\theta,n+1}$ for $n \in \mathbb{N}$ and $(\overline{b}_1 = i) = I_\theta(i) \times [0, \theta], i \in \mathbb{N}_m$, it follows that

$$\overline{\gamma}_\theta\left(\overline{b}_1 = i \mid \overline{b}_0, \overline{b}_{-1}, \ldots\right) = \lim_{n \to \infty} \overline{\gamma}_\theta\left(\overline{b}_1 = i \mid \overline{b}_0, \overline{b}_{-1}, \ldots, \overline{b}_{-n}\right)$$

$$= \lim_{n \to \infty} \overline{\gamma}_\theta\left(\overline{b}_1 = i \mid [0, \theta] \times I_{\theta,n+1}\right) = \lim_{n \to \infty} \frac{\overline{\gamma}_\theta\left(I_\theta(i) \times I_{\theta,n+1}\right)}{\overline{\gamma}_\theta\left([0, \theta] \times I_{\theta,n+1}\right)}$$

$$= \lim_{n \to \infty} \frac{1}{\overline{\gamma}_\theta(I_{\theta,n+1})} \int_{I_{\theta,n+1}} P_{\theta,i}(y) d\overline{\gamma}_\theta(y) = \lim_{n \to \infty} P_{\theta,i}(y_{n+1}) = P_{\theta,i}(b) \quad \overline{\gamma}_\theta\text{-a.s.},$$

(2.4.12)

where $y_{n+1} \in I_{\theta,n+1}$ with $\lim_{n \to \infty} y_{n+1} = [\overline{b}_0 \theta, \overline{b}_{-1} \theta, \ldots] = b$ and $P_{\theta,i}$ is as in (2.2.20). The strict stationarity of $(\overline{b}_l)_{l \in \mathbb{Z}}$ under $\overline{\gamma}_\theta$ implies that

$$\overline{\gamma}_\theta(\overline{b}_{l+1} = i \mid \overline{b}_l, \overline{b}_{l-1}, \ldots) = P_{\theta,i}(b) \quad \overline{\gamma}_\theta\text{-a.s.} \quad (2.4.13)$$

for any $i \in \mathbb{N}_m$ and $l \in \mathbb{Z}$. The last equation emphasizes that $(\overline{b}_l)_{l \in \mathbb{Z}}$ is an infinite order chain in the theory of dependence with complete connections (see [28], Section 5.5).

Motivated by (2.4.11), we shall consider the one-parameter family $\{\gamma_{\theta,b} : b \in [0, \theta]\}$ of (conditional) probability measures on $([0, \theta], \mathcal{B}_{[0,\theta]})$ defined by their distribution functions

$$\gamma_{\theta,b}([0, x]) := \frac{(b\theta + 1)x}{(bx + 1)\theta}, \quad x \in [0, \theta], \ b \in [0, \theta]. \quad (2.4.14)$$

Note that $\gamma_{\theta,0}$ is the Lebesgue measure λ_θ on $[0, \theta]$.

Let b_n's be as in (2.1.4). For each $b \in [0, \theta]$, define $(s_{\theta,n}^b)_{n \in \mathbb{N}_+}$ by

$$s_{\theta,0}^b := b, \quad s_{\theta,n}^b := \frac{1}{b_n \theta + s_{\theta,n-1}}, \quad n \in \mathbb{N}_+.$$

Then we have $s_{\theta,1}^b = 1/(b_1 \theta + b)$, $s_{\theta,n}^b = [b_n \theta, \ldots, b_2 \theta, b_1 \theta + b]$, $n \geq 2$. Note that $\gamma_{\theta,b}(A | b_1, \ldots, b_n) = \gamma_{\theta, s_{\theta,n}^b}(T_\theta^n(A))$, for all $b \in [0, \theta], n \in \mathbb{N}_+$ and for any A belonging to the σ-algebra generated by the random variables $b_{n+1}, b_{n+2}, \ldots$, that is, $T_\theta^{-n}(\mathcal{B}_{[0,\theta]})$. In particular, it follows that for any $b \in [0, \theta]$

$$\gamma_{\theta,b}\left(T_\theta^n < x \mid b_1, \ldots, b_n\right) = \frac{(s_{\theta,n}^b \theta + 1)x}{(s_{\theta,n}^b x + 1)\theta} \quad (2.4.15)$$

for any $x \in [0, \theta], n \in \mathbb{N}_+$.

2.5 The Perron-Frobenius Operator of T_θ Under γ_θ

Let $([0, \theta], \mathcal{B}_{[0,\theta]}, \gamma_\theta, T_\theta)$ be as in Definition 2.1.1. In this section, we derive its Perron-Frobenius operator.

Let μ be a probability measure on $([0, \theta], \mathcal{B}_{[0,\theta]})$ such that $\mu\left(T_\theta^{-1}(A)\right) = 0$ whenever $\mu(A) = 0$ for $A \in \mathcal{B}_{[0,\theta]}$. This condition is satisfied if T_θ is μ-preserving, that is, $\mu T_\theta^{-1} = \mu$. Let $L^1([0, \theta], \mu) := \{f : [0, \theta] \to \mathbb{C} : \int_0^\theta |f|\,d\mu < \infty\}$. The *Perron-Frobenius operator* P_μ of $([0, \theta], \mathcal{B}_{[0,\theta]}, \mu, T_\theta)$ is defined as the bounded linear operator on the Banach space $L^1([0, \theta], \mu)$ such that the following holds:

$$\int_A P_\mu f\,d\mu = \int_{T_\theta^{-1}(A)} f\,d\mu \quad \text{for all } A \in \mathcal{B}_{[0,\theta]},\ f \in L^1([0, \theta], \mu). \tag{2.5.1}$$

Proposition 2.5.1 ([64]) *Let $([0, \theta], \mathcal{B}_{[0,\theta]}, \gamma_\theta, T_\theta)$ be as in Definition 2.1.1:*

(i) *The Perron-Frobenius operator $U_\theta := P_{\gamma_\theta}$ of T_θ under the invariant measure γ_θ on $\mathcal{B}_{[0,\theta]}$ is given a.e. in $[0, \theta]$ by the equation:*

$$U_\theta f(x) = \sum_{i \geq m} P_{\theta,i}(x)\, f(u_{\theta,i}(x)), \quad m \in \mathbb{N}_+,\ f \in L^1([0, \theta], \gamma_\theta), \tag{2.5.2}$$

where $P_{\theta,i}$ and $u_{\theta,i}$, $i \geq m$, are as in (2.2.20) and (2.4.1), respectively.

(ii) *The Perron-Frobenius operator $S_\theta := P_{\lambda_\theta}$ of T_θ under the Lebesgue measure λ_θ on $\mathcal{B}_{[0,\theta]}$ is given a.e. in $[0, \theta]$ by the equation*

$$S_\theta f(x) = \sum_{i \geq m} \frac{1}{(i\theta + x)^2} f(u_{\theta,i}(x)),\ f \in L^1([0, \theta], \lambda_\theta). \tag{2.5.3}$$

The powers of S_θ are given a.e. in $[0, \theta]$ by the equation

$$S_\theta^n f(x) = \frac{U_\theta^n g_\theta(x)}{1 + \theta x},\ f \in L^1([0, \theta], \lambda_\theta),\ n \in \mathbb{N}_+, \tag{2.5.4}$$

where $g_\theta(x) := (1 + \theta x) f(x)$, $x \in [0, \theta]$.

(iii) *Let μ be a probability measure on $\mathcal{B}_{[0,\theta]}$. Assume that $\mu \ll \lambda_\theta$ and let $h_\theta = d\mu/d\lambda_\theta$ a.e. in $[0, \theta]$. For any $n \in \mathbb{N}_+$ and $A \in \mathcal{B}_{[0,\theta]}$, we have*

$$\mu\left(T_\theta^{-n}(A)\right) = \int_A U_\theta^n f_\theta(x)\,d\gamma_\theta(x), \tag{2.5.5}$$

where $f_\theta(x) := (\log(1 + \theta^2))\frac{1+\theta x}{\theta^2} h_\theta(x)$, $x \in [0, \theta]$.

Let $B([0,\theta])$ denote the collection of all bounded measurable functions $f : [0,\theta] \to \mathbb{C}$. A different interpretation is available for the operator U_θ restricted to $B([0,\theta])$, which is a Banach space under the supremum norm.

Proposition 2.5.2 ([64]) *The operator $U_\theta : B([0,\theta]) \to B([0,\theta])$ is the transition operator of the Markov chain $(s_{\theta,n}^b)_{n\in\mathbb{N}_+}$ on $([0,\theta], \mathcal{B}_{[0,\theta]}, \gamma_{\theta,b})$, for any $b \in [0,\theta]$.*

In the following proposition we show that the operator $U_\theta : B([0,\theta]) \to B([0,\theta])$ in (2.5.2) preserves monotonicity and enjoys a contraction property for Lipschitz continuous functions.

Proposition 2.5.3 ([41]) *Let U_θ be as in (2.5.2):*

(i) Let $f \in B([0,\theta])$. Then the following holds:
  (a) If f is nondecreasing (nonincreasing), then $U_\theta f$ is nonincreasing (nondecreasing).
  (b) If f is monotone, then

$$\text{var}\,(U_\theta f) \leq \frac{1}{m+1} \cdot \text{var}\, f. \tag{2.5.6}$$

(ii) For any $f \in L([0,\theta])$, we have

$$s(U_\theta f) \leq \widetilde{q}_\theta \cdot s(f), \tag{2.5.7}$$

where

$$\widetilde{q}_\theta := m \cdot \sum_{i \geq m} \left(\frac{m}{i^3(i+1)} + \frac{i+1-m}{i(i+1)^3} \right). \tag{2.5.8}$$

Remark 2.5.4 In hypothesis of Proposition 2.5.1(iii) it follows that

$$\mu(T_\theta^{-n}(A)) - \gamma_\theta(A) = \int_A (U_\theta^n f_\theta(x) - 1)\mathrm{d}\gamma_\theta(x), \tag{2.5.9}$$

for any $n \in \mathbb{N}$ and $A \in \mathcal{B}_{[0,\theta]}$, where $f_\theta(x) := \left(\log(1+\theta^2)\right) \frac{1+\theta x}{\theta^2} h_\theta(x)$, $x \in [0,\theta]$. The last equation shows that the asymptotic behavior of $\mu(T_\theta^{-n}(A)) - \gamma_\theta(A)$ as $n \to \infty$ is given by the asymptotic behavior of the n-th power of the Perron-Frobenius U_θ on $L^1([0,\theta], \gamma_\theta)$ or on smaller Banach spaces.

2.5 The Perron-Frobenius Operator of T_θ Under γ_θ

Let $BV([0, \theta]) := \{f : [0, \theta] \to \mathbb{C} : \text{var } f < \infty\}$. It is known that $BV([0, \theta]) \subset B([0, \theta]) \subset L^1([0, \theta])$. If $f \in B([0, \theta])$ define the linear functional U_θ^∞ by

$$U_\theta^\infty : B([0, \theta]) \to \mathbb{C}; \quad U_\theta^\infty f = \int_0^\theta f(x) \gamma_\theta(dx), \tag{2.5.10}$$

then we have $U_\theta^\infty U_\theta^n f = U_\theta^\infty f$ for any $n \in \mathbb{N}_+$.

Proposition 2.5.5 ([65]) *For any $f \in BV([0, \theta])$ and for all $n \in \mathbb{N}_+$, we have*

$$\text{var } U_\theta^n f \leq \frac{1}{(m+1)^n} \text{var } f, \tag{2.5.11}$$

$$\left| U_\theta^n f - U_\theta^\infty f \right| \leq \frac{1}{(m+1)^n} \text{var } f. \tag{2.5.12}$$

By induction with respect to $n \in \mathbb{N}$ we get

$$U_\theta^n f(x) = \sum_{i_1, \dots, i_n \in \mathbb{N}_m} P_{\theta, i_1 \dots i_n}(x) f(u_{\theta, i_n \dots i_1}(x)), \quad x \in [0, \theta],$$

where

$$u_{\theta, i_n \dots i_1} := u_{\theta, i_n} \circ \dots \circ u_{\theta, i_1} \tag{2.5.13}$$

$$P_{\theta, i_1 \dots i_n}(x) := P_{\theta, i_1}(x) P_{\theta, i_2}(u_{\theta, i_1}(x)) \dots P_{\theta, i_n}(u_{\theta, i_{n-1} \dots i_1}(x)), \quad n \geq 2. \tag{2.5.14}$$

Here the functions $u_{\theta, i}$ and $P_{\theta, i}$ are defined in (2.4.1) and (2.2.20), respectively, for all $i \in \mathbb{N}_m$. Putting

$$\frac{p_n(i_1, \dots, i_n)}{q_n(i_1, \dots, i_n)} = [i_1\theta, \dots, i_n\theta], \quad n \in \mathbb{N}_+,$$

for arbitrary indeterminates $i_1, \dots, i_n$, we get

$$P_{\theta, i_1 \dots i_n}(b) = \frac{1 + b\theta}{q_{n-1}(i_2, \dots, i_n)(b + i_1\theta) + p_{n-1}(i_2, \dots, i_n)}$$

$$\times \frac{1}{q_n(i_2, \dots, i_n, m)(b + i_1\theta) + p_n(i_2, \dots, i_n, m)} \tag{2.5.15}$$

for all $n \geq 2$, $i_n \in \mathbb{N}_m$, and $b \in [0, \theta]$.

2.6 Ergodicity of the Associated RSCC

The facts presented above lead us to a certain random system with complete connections (RSCC) associated with the θ-expansion. Using the properties presented in Sect. 1.4, we are able to study the following RSCC

$$\left\{ ([0,\theta], \mathcal{B}_{[0,\theta]}), (\mathbb{N}_m, \mathcal{P}(\mathbb{N}_m)), u_\theta, P_\theta \right\}, \tag{2.6.1}$$

where $u_\theta(s,i) := u_{\theta,i}(s)$ and $P_\theta(s,i) := P_{\theta,i}(s)$, $s \in [0,\theta]$, $i \in \mathbb{N}_m$, are given in (2.4.1) and (2.2.20). Here $\mathbb{N}_m := \{m, m+1, \ldots\}$, $m \in \mathbb{N}_+$ and $\mathcal{P}(\mathbb{N}_m)$ denotes the power set of $\mathbb{N}_m$.

Whatever $b \in [0,\theta]$ the Markov chain $\left(s_{\theta,n}^b \right)_{n \in \mathbb{N}}$ associated with the RSCC (2.6.1) has the transition operator U_θ, with the transition probability function

$$Q_\theta(s, B) = \sum_{\{i \geq m \mid u_{\theta,i}(s) \in B\}} P_{\theta,i}(s), \quad s \in [0,\theta], B \in \mathcal{B}_{[0,\theta]}. \tag{2.6.2}$$

Then $Q_\theta^n(\cdot, \cdot)$ will denote the n-step transition probability function of the same Markov chain.

Proposition 2.6.1 ([64]) *The RSCC (2.6.1) is regular with respect to $L([0,\theta])$. Moreover there exist a stationary probability measure $Q_\theta^\infty = \gamma_\theta$ and two positive constants $q_\theta < 1$ and K_θ such that*

$$\left\| U_\theta^n f - \int_0^\theta f \, d\gamma_\theta \right\|_L \leq K_\theta q_\theta^n \|f\|_L, \quad n \in \mathbb{N}_+, f \in L([0,\theta]), \tag{2.6.3}$$

where

$$U_\theta^n f(\cdot) := \int_0^\theta Q_\theta^n(\cdot, ds) f(s). \tag{2.6.4}$$

Proof Since

$$\frac{d}{ds} u_{\theta,i}(s) = \frac{-1}{(s+i\theta)^2}, \quad \frac{d}{ds} P_{\theta,i}(s) = \frac{i\theta - 1/\theta}{(s+i\theta)^2} - \frac{(i+1)\theta - 1/\theta}{(s+(i+1)\theta)^2},$$

for any $s \in [0,\theta]$ and $i \geq m$, it follows that

$$\sup_{s \in [0,\theta]} \left| \frac{d}{ds} u_{\theta,i}(s) \right| = \frac{1}{(i\theta)^2} \text{ and } \sup_{s \in [0,\theta]} \left| \frac{d}{ds} P_{\theta,i}(s) \right| < \infty.$$

2.6 Ergodicity of the Associated RSCC

Hence the requirements of Definition 1.4.8 of an RSCC with contraction are met with $l = 1$. By Theorem 1.4.5 it follows that the Markov chain $(s^b_{\theta,n})_{n \in \mathbb{N}}$ associated with this RSCC with contraction is a Doeblin-Fortet chain and its transition operator U_θ is a Doeblin-Fortet operator. It remains to prove the regularity of U_θ with respect to $L([0, \theta])$. For this we have to prove the existence of a point $s^* \in [0, \theta]$ such that $\lim_{n \to \infty} |\sigma_n(s) - s^*| = 0$, for any $s \in [0, \theta]$, where $\sigma_n(s)$ is the support of measure $Q^n_\theta(s, \cdot)$, $n \in \mathbb{N}_+$.

Let $s \in [0, \theta]$ be an arbitrarily fixed number and define

$$w_1 := s, \quad w_{n+1} := \frac{1}{w_n + m\theta}, \quad n \in \mathbb{N}_+. \tag{2.6.5}$$

We have $w_n \in [0, \theta]$, and letting $n \to \infty$ in (2.6.5), we get

$$w_n \to s^* := \frac{-1 + \sqrt{1 + 4\theta^2}}{2\theta}.$$

Clearly, $w_{n+1} \in \sigma_1(w_n)$ and Theorem 1.4.9(ii) and an induction argument show that $w_n \in \sigma_n(s)$, $n \in \mathbb{N}_+$. Thus $d(\sigma_n(s), s^*) \leq |w_n - s^*| \to 0$, $n \to \infty$ where d stands for the Euclidian distance on the line. Now, the regularity of U_θ with respect to $L([0, \theta])$ follows from Theorem 1.4.9(i). From (1.4.12) and Theorem 1.4.6(ii) there exist a unique stationary probability measure Q^∞_θ and two constants $q_\theta < 1$ and K_θ such that

$$\|U^n_\theta f - U^\infty_\theta f\|_L \leq K_\theta q^n_\theta \|f\|_L, \quad n \in \mathbb{N}_+, \ f \in L([0, \theta]), \tag{2.6.6}$$

where $U^n_\theta f$ is as in (2.6.4) and

$$U^\infty_\theta f = \int_0^\theta f(x) Q^\infty_\theta(dx). \tag{2.6.7}$$

Since

$$\int_0^\theta Q_\theta(x, B) \gamma_\theta(dx) = \gamma_\theta(B), \quad x \in [0, \theta], \ B \in \mathcal{B}_{[0,\theta]}$$

on account of the uniqueness of Q^∞_θ, it follows that Q^∞_θ coincides with the invariant probability measure of the transformation T_θ in (2.1.2), i.e., Q^∞_θ has the density $\rho_\theta(x) = 1/(x + m\theta)$, $x \in [0, \theta]$, with the normalizing factor $1/\log(1 + \theta^2)$. □

Remark 2.6.2 Another way to put this is that ρ_θ is the eigenfunction of eigenvalue 1 of the Perron-Frobenius operator S_θ under the Lebesgue measure λ_θ.

2.7 A Gauss-Kuzmin-Type Theorem for T_θ

Let μ be a nonatomic probability measure on $\mathcal{B}_{[0,\theta]}$ and define

$$F_{\theta,n}(x) := \mu(T_\theta^n \leq x), \quad x \in [0,\theta], \ n \in \mathbb{N}, \qquad (2.7.1)$$

$$F_\theta(x) := \lim_{n \to \infty} F_{\theta,n}(x), \quad x \in [0,\theta], \qquad (2.7.2)$$

with $F_{\theta,0}(x) = \mu([0,x])$. Now, we may determine the limit of the sequence $(F_{\theta,n})_{n \in \mathbb{N}}$ as $n \to \infty$ and give the rate of this convergence.

Theorem 2.7.1 ([64]) *Let $([0,\theta], \mathcal{B}_{[0,\theta]}, T_\theta)$ be as in Definition 2.1.1 and F_θ as in (2.7.2):*

(i) For a probability measure μ on $([0,\theta], \mathcal{B}_{[0,\theta]})$, let assumption (A) as follows:

(A) μ *is nonatomic and has a Riemann-integrable density.*

Then for any probability measure μ which satisfies (A), the following holds:

$$F_\theta(x) = \frac{1}{\log(1+\theta^2)} \log(1+\theta x), \quad x \in [0,\theta]. \qquad (2.7.3)$$

(ii) In addition to assumption of μ in (i), if the density of $[0,\theta] \ni x \mapsto \mu([0,x])$ is Lipschitz continuous, then there exist two positive constants $q_\theta < 1$ and K_θ such that for any $x \in [0,\theta]$ and $n \in \mathbb{N}_+$, the following holds:

$$F_{\theta,n}(x) = \frac{1 + \alpha_\theta q_\theta^n}{\log(1+\theta^2)} \log(1+\theta x), \qquad (2.7.4)$$

where $\alpha_\theta := \alpha_\theta(\mu, n, x)$ with $|\alpha_\theta| \leq K_\theta$. As a consequence, the n-th error term $e_{\theta,n}(\mu; x)$ of the Gauss-Kuzmin problem is obtained as follows:

$$e_{\theta,n}(\mu; x) = \frac{\alpha_\theta q_\theta^n}{\log(1+\theta^2)} \log(1+\theta x). \qquad (2.7.5)$$

Proof Let T_θ be as in (2.1.2). By Proposition 2.5.1(iii), we have

$$\mu\left(T_\theta^{-n}(A)\right) = \int_A U_\theta^n f_{\theta,0}(x) \rho_\theta(x) dx \quad \text{for any } n \in \mathbb{N}, \ A \in \mathcal{B}_{[0,\theta]}, \qquad (2.7.6)$$

where $f_{\theta,0}(x) = \frac{1+\theta x}{\theta^2}(\mathrm{d}\mu/\mathrm{d}\lambda_\theta)(x)$ for $x \in [0,\theta]$. If $\mathrm{d}\mu/\mathrm{d}\lambda_\theta \in L([0,\theta])$, by (2.6.7) we have

$$U_\theta^\infty f_{\theta,0} = \int_0^\theta f_{\theta,0}(x)\, Q_\theta^\infty(\mathrm{d}x) = \int_0^\theta f_{\theta,0}(x)\, \gamma_\theta(\mathrm{d}x) = \frac{1}{\log(1+\theta^2)}. \quad (2.7.7)$$

Taking into account (2.6.6), there exist two constants $q_\theta < 1$ and K_θ such that

$$\|U_\theta^n f - U_\theta^\infty f\|_L \le K_\theta q_\theta^n \|f\|_L, \quad n \in \mathbb{N}_+,\ f \in L([0,\theta]).$$

Hence, $\|U_\theta^n f - U_\theta^\infty f\|_L = \mathcal{O}(q_\theta^n) \|f\|_L$ as $n \to \infty$ for some positive $q_\theta < 1$, the constant implied in $\mathcal{O}$ being independent of $f \in L([0,\theta])$. Furthermore, consider the Banach space $C([0,\theta])$ of all real-valued continuous functions on $[0,\theta]$ with the norm $\|f\| := \sup_{x\in[0,\theta]} |f(x)|$. Since $L([0,\theta])$ is a dense subspace of $C([0,\theta])$, we have

$$\lim_{n\to\infty} \|U_\theta^n f - U_\theta^\infty f\| = 0 \quad \text{for all } f \in C([0,\theta]). \quad (2.7.8)$$

Therefore, (2.7.8) implies that $\lim_{n\to\infty} |U_\theta^n f(x) - U_\theta^\infty f(x)| = 0$ for all $x \in [0,\theta]$ and any measurable function f which is Q_θ^∞-almost surely continuous, i.e., for any Riemann-integrable function on $[0,\theta]$. Thus, we have

$$F_\theta(x) = \lim_{n\to\infty} F_{\theta,n}(x) = \lim_{n\to\infty} \mu\left(T_\theta^n \le x\right) = \lim_{n\to\infty} \int_0^x U_\theta^n f_{\theta,0}(u) \rho_\theta(u)\, \mathrm{d}u$$

$$= \frac{1}{\log(1+\theta^2)} \int_0^x \rho_\theta(u)\, \mathrm{d}u = \frac{1}{\log(1+\theta^2)} \log(1+\theta x).$$

$\square$

Remark 2.7.2 Since the Lebesgue measure λ_θ satisfies assumptions in both (i) and (ii) of Theorem 2.7.1, (2.7.3) and (2.7.4) hold for the case $\mu = \lambda_\theta$. Hence Theorem 2.7.1 gives the solution of the Gauss-Kuzmin problem for the pair (T_θ, μ).

2.8 Szüsz's Method to Gauss-Kuzmin-Type Theorem

Applying Szüsz's method [73] we obtain more information on the convergence rate than that obtained by applying the method of dependence with complete connections. We have the following results.

Theorem 2.8.1 *Let T_θ and $F_{\theta,n}$ be as in (2.1.2) and (2.7.1), respectively. Then there exists a constant $0 < \widetilde{q}_\theta < 1$ such that $F_{\theta,n}$ can be written as*

$$F_{\theta,n}(x) = \frac{\log(1+\theta x)}{\log(1+\theta^2)} + \mathcal{O}((\widetilde{q}_\theta)^n) \tag{2.8.1}$$

uniformly with respect to $x \in [0, \theta]$.

Remark 2.8.2 From (2.8.1), we see that

$$F_\theta(x) = \lim_{n\to\infty} F_{\theta,n}(x) = \gamma_\theta([0,x]). \tag{2.8.2}$$

To prove Theorem 2.8.1 we need the following results. First, we show that $\{F_{\theta,n}\}$ in (2.7.1) satisfy a Gauss-Kuzmin-type equation.

Lemma 2.8.3 *For the distribution functions $\{F_{\theta,n}\}$ in (2.7.1), the following Gauss-Kuzmin-type equation holds:*

$$F_{\theta,n+1}(x) = \sum_{i \geq m} \left\{ F_{\theta,n}\left(\frac{1}{i\theta}\right) - F_{\theta,n}\left(\frac{1}{i\theta+x}\right) \right\} \tag{2.8.3}$$

for $x \in [0, \theta]$ and $n \in \mathbb{N}$.

Proof Let $I_{\theta,n} = \{x \in [0, \theta] : T_\theta^n(x) \leq x\}$ and

$$I_{\theta,n,i} = \left\{ x \in I_{\theta,n} : \frac{1}{i\theta + x} < T_\theta^n(x) < \frac{1}{i\theta} \right\}. \tag{2.8.4}$$

From (2.1.2) and (2.1.4), we see that

$$T_\theta^n(x) = \frac{1}{b_{n+1}\theta + T_\theta^{n+1}(x)}, \quad n \in \mathbb{N}_+. \tag{2.8.5}$$

From the definition of $I_{\theta,n,i}$ and (2.8.5) it follows that for any $n \in \mathbb{N}$, $I_{\theta,n+1} = \bigcup_{i \geq m} I_{\theta,n,i}$. Then (2.8.3) holds because $F_{\theta,n+1}(x) = \mu(I_{\theta,n+1})$ and

$$\mu(I_{\theta,n,i}) = F_{\theta,n}\left(\frac{1}{i\theta}\right) - F_{\theta,n}\left(\frac{1}{i\theta+x}\right). \tag{2.8.6}$$

□

2.8 Szüsz's Method to Gauss-Kuzmin-Type Theorem

Remark 2.8.4 Suppose that $F'_{\theta,0}$ exists everywhere in $[0,\theta]$ and is bounded (μ has bounded density). Then by induction we have that $F'_{\theta,n}$ exists and it is bounded for any $n \in \mathbb{N}_+$. This allows us to differentiate (2.8.3) term by term, obtaining

$$F'_{\theta,n+1}(x) = \sum_{i \geq m} \frac{1}{(i\theta + x)^2} F'_{\theta,n}\left(\frac{1}{i\theta + x}\right). \tag{2.8.7}$$

We introduce functions $\{f_{\theta,n}\}$ as follows:

$$f_{\theta,n}(x) := (1 + \theta x) F'_{\theta,n}(x), \quad x \in [0,\theta], \, n \in \mathbb{N}. \tag{2.8.8}$$

Then (2.8.7) is

$$f_{\theta,n+1}(x) = \sum_{i \geq m} P_{\theta,i}(x) f_{\theta,n}\left(u_{\theta,i}(x)\right), \tag{2.8.9}$$

where $P_{\theta,i}(x)$ and $u_{\theta,i}(x)$ are given in (2.2.20) and (2.4.1), respectively. By Proposition 2.5.1 (i), we have that $f_{\theta,n+1}(x) = U_\theta f_{\theta,n}(x)$.

Lemma 2.8.5 *For $\{f_{\theta,n}\}$ in (2.8.8), define $M_{\theta,n} := \max_{x \in [0,\theta]} |f'_{\theta,n}(x)|$. Then*

$$M_{\theta,n+1} \leq \widetilde{q}_\theta \cdot M_{\theta,n}, \tag{2.8.10}$$

where $\widetilde{q}_\theta$ is the constant in (2.5.8).

Proof Since

$$P_{\theta,i}(x) = \frac{1}{\theta}\left(\frac{1 - i\theta^2}{x + i\theta} - \frac{1 - (i+1)\theta^2}{x + (i+1)\theta}\right), \tag{2.8.11}$$

we have

$$f'_{\theta,n+1}(x) = \sum_{i \geq m} \frac{1 - (i+1)\theta^2}{(x + i\theta)(x + (i+1)\theta)^3} f'_{\theta,n}(\alpha_{\theta,i}) - \sum_{i \geq m} \frac{P_{\theta,i}(x)}{(x + i\theta)^2} f'_{\theta,n}(u_{\theta,i}(x)), \tag{2.8.12}$$

where $u_{\theta,i+1}(x) < \alpha_{\theta,i} < u_{\theta,i}(x)$. Now (2.8.12) implies

$$M_{\theta,n+1} \leq M_{\theta,n} \cdot \max_{x \in [0,\theta]} \left(\sum_{i \geq m} \frac{(i+1)\theta^2 - 1}{(x + i\theta)(x + (i+1)\theta)^3} + \sum_{i \geq m} \frac{P_{\theta,i}(x)}{(x + i\theta)^2}\right). \tag{2.8.13}$$

We now must calculate the maximum value of the sums in this expression. First, we note that

$$\frac{(i+1)\theta^2 - 1}{(x+i\theta)(x+(i+1)\theta)^3} \leq m^2 \frac{(i+1)\theta^2 - 1}{i(i+1)^3}, \tag{2.8.14}$$

where we used that $0 \leq x \leq \theta$. Next, let

$$\widetilde{h}_\theta(x) := \sum_{i \geq m} \frac{P_{\theta,i}(x)}{(x+i\theta)^2}. \tag{2.8.15}$$

By Proposition 2.5.1 (i) and Proposition 2.5.3 (i)(a), we have that the function $\widetilde{h}_\theta$ is decreasing for $x \in [0, \theta]$. Hence, $\widetilde{h}_\theta(x) \leq \widetilde{h}_\theta(0)$. This leads to

$$\sum_{i \geq m} \frac{P_{\theta,i}(x)}{(x+i\theta)^2} \leq m^2 \cdot \sum_{i \geq m} \frac{1}{i^3(i+1)}. \tag{2.8.16}$$

The relations (2.8.13), (2.8.14), and (2.8.16) imply (2.8.10) and that $\widetilde{q}_\theta$ is as in (2.5.8). □

Proof of Theorem 2.8.1 For $\{F_{\theta,n}\}$ in (2.7.1), we introduce a function $R_{\theta,n}(x)$ such that

$$F_{\theta,n}(x) = \frac{\log(1+\theta x)}{\log(1+\theta^2)} + R_{\theta,n}(x). \tag{2.8.17}$$

Because $F_{\theta,n}(0) = 0$ and $F_{\theta,n}(\theta) = 1$, we have $R_{\theta,n}(0) = R_{\theta,n}(\theta) = 0$. To prove Theorem 2.8.1, we have to show the existence of a constant $0 < \widetilde{q}_\theta < 1$ such that

$$R_{\theta,n}(x) = \mathcal{O}((\widetilde{q}_\theta)^n). \tag{2.8.18}$$

For $\{f_{\theta,n}\}$ in (2.8.8), if we can show that $f_{\theta,n}(x) = \dfrac{\theta}{\log(1+\theta^2)} + \mathcal{O}((\widetilde{q}_\theta)^n)$, then its integration will show Eq. (2.7.3). □

To demonstrate that $f_{\theta,n}(x)$ has this desired form, it suffices to prove the following lemma.

Lemma 2.8.6 *For any $x \in [0, \theta]$ there exists a constant $\widetilde{q}_\theta$ with $0 < \widetilde{q}_\theta < 1$ such that*

$$f'_{\theta,n}(x) = \mathcal{O}((\widetilde{q}_\theta)^n). \tag{2.8.19}$$

Proof Let $\widetilde{q}_\theta$ be as in (2.5.8). Using Lemma 2.8.5, to show (2.8.19) it is enough to prove that $\widetilde{q}_\theta < 1$ (Table 2.1).

Table 2.1 Error values for some $m \geq 1$

m	θ	$\widetilde{q}_\theta$
1	1	0.7591797
2	0.7071067	0.3265870
3	0.5773502	0.2022220
10	0.3162277	0.0533201
20	0.2236067	0.0258325
30	0.1825741	0.0170368
40	0.1581138	0.0127082
50	0.1414213	0.0101333
100	0.1000000	0.0050333
1000	0.0316227	0.0005003
5000	0.0141421	0.0001000

We observe that for various values of m, we have $\widetilde{q}_\theta < 1$. The more m increases, the more $\widetilde{q}_\theta$ decreases. We note that for the case $m = 1$, that is, in the RCF case, we obtain $\widetilde{q}_\theta = 0.7591797$, value obtained by P. Szüsz [73]. For $m \geq 2$ the previous estimates are not too bad, but they can be improved (see Appendix 2.9.1). □

2.9 A Near-Optimal Solution to the Gauss-Kuzmin-Lévy Problem

An operatorial treatment on suitable Banach spaces allowed us to study the optimality of the convergence rate. Actually, in [63] we obtained upper and lower bounds of the convergence rate which provide a near-optimal solution to the Gauss-Kuzmin-Lévy problem.

Let μ be a probability measure on $\mathcal{B}_{[0,\theta]}$ such that $\mu \ll \lambda_\theta$. For any $n \in \mathbb{N}$ put $F_{\theta,n}(x) = \mu(T_\theta^n < x)$, $x \in [0, \theta]$, where T_θ^0 is the identity map. As $(T_\theta^n < x) = T_\theta^{-n}((0, x))$, by Proposition 3.4.1 (iii), we have

$$F_{\theta,n}(x) = \int_0^x \frac{U_\theta^n f_{\theta,0}(u)}{1 + \theta u} \theta \, du, \quad n \in \mathbb{N}, \tag{2.9.1}$$

with $f_{\theta,0}(x) := \frac{1+\theta x}{\theta^2}(F_{\theta,0})'(x)$, $x \in [0, \theta]$, where $(F_{\theta,0})' = d\mu/d\lambda_\theta$.

We will assume that $(F_{\theta,0})' \in C^1([0, \theta])$. So, we study the behavior of U_θ^n as $n \to \infty$, assuming that the domain of U_θ is $C^1([0, \theta])$, the collection of all functions $f : [0, \theta] \to \mathbb{C}$ which have a continuous derivative.

Let $f \in C^1([0, \theta])$. Then the series (2.5.2) can be differentiated term by term, since the series of derivatives is uniformly convergent. We get

$$(U_\theta f)'(x) = \sum_{j \geq m} \left\{ (P_{\theta,j}(x))' f\left(\frac{1}{x+j\theta}\right) - P_{\theta,j}(x) f'\left(\frac{1}{x+j\theta}\right) \frac{1}{(x+j\theta)^2} \right\}$$

$$= \sum_{j \geq m} \left\{ \left(\frac{j\theta - \frac{1}{\theta}}{(x+j\theta)^2} - \frac{(j+1)\theta - \frac{1}{\theta}}{(x+(j+1)\theta)^2} \right) f\left(\frac{1}{x+j\theta}\right) \right.$$

$$\left. - P_{\theta,j}(x) \frac{1}{(x+j\theta)^2} f'\left(\frac{1}{x+j\theta}\right) \right\}$$

$$= -\sum_{j \geq m} \left\{ \frac{(j+1)\theta - \frac{1}{\theta}}{(x+(j+1)\theta)^2} \left(f\left(\frac{1}{x+j\theta}\right) - f\left(\frac{1}{x+(j+1)\theta}\right) \right) \right.$$

$$\left. + P_{\theta,j}(x) \frac{1}{(x+j\theta)^2} f'\left(\frac{1}{x+j\theta}\right) \right\}, \quad x \in [0,\theta]. \tag{2.9.2}$$

Thus, we can write

$$(U_\theta f)' = -U_\theta^* f', \quad f \in C^1([0,\theta]), \tag{2.9.3}$$

where $U_\theta^* : C([0,\theta]) \to C([0,\theta])$ is defined by

$$U_\theta^* g(x) = \sum_{j \geq m} \left\{ \frac{(j+1)\theta - \frac{1}{\theta}}{(x+(j+1)\theta)^2} \int_{\frac{1}{x+(j+1)\theta}}^{\frac{1}{x+j\theta}} g(u) du + P_{\theta,j}(x) \frac{1}{(x+j\theta)^2} g\left(\frac{1}{x+j\theta}\right) \right\} \tag{2.9.4}$$

with $g \in C([0,\theta))$ and $x \in [0,\theta]$. Clearly, $(U_\theta^n f)' = (-1)^n (U_\theta^*)^n f'$, $n \in \mathbb{N}_+$, $f \in C^1([0,\theta])$.

We are going to show that $(U_\theta^*)^n$ takes certain functions into functions with very small values when $n \in \mathbb{N}_+$ is large.

Proposition 2.9.1 ([63]) *There are positive constants $v_\theta < w_\theta < 1$ and a real-valued function $\varphi_\theta \in C([0,\theta])$ such that*

$$v_\theta \varphi_\theta \leq U_\theta^* \varphi_\theta \leq w_\theta \varphi_\theta, \quad \theta^2 = 1/m, \, m \in \mathbb{N}_+. \tag{2.9.5}$$

Proof Let $h_\theta^* : \mathbb{R}_+ \to \mathbb{R}$, with $\theta^2 = 1/m$, $m \in \mathbb{N}_+$, be a continuous bounded function such that $\lim_{x \to \infty} h_\theta^*(x) < \infty$. We look for a function $g_\theta^* : (0,\theta] \to \mathbb{R}$ such that $U_\theta g_\theta^* = h_\theta^*$, assuming that the equation

$$U_\theta g_\theta^*(x) = \sum_{j \geq m} P_{\theta,j}(x) g_\theta^*\left(u_j(x)\right) = h_\theta^*(x) \tag{2.9.6}$$

2.9 A Near-Optimal Solution to the Gauss-Kuzmin-Lévy Problem

holds for $x \in \mathbb{R}_+$. By reducing the terms of the series involved, (2.9.6) yields

$$\frac{h_\theta^*(x)}{1+\theta x} - \frac{h_\theta^*(\theta + x)}{1+\theta(x+\theta)} = \frac{1}{(x+m\theta)(x+(m+1)\theta)} g_\theta^*\left(\frac{1}{x+m\theta}\right), \quad x \in \mathbb{R}_+.$$
(2.9.7)

Hence

$$g_\theta^*(u) = \left(\frac{1}{u\theta} + 1\right) h_\theta^*\left(\frac{1}{u} - m\theta\right) - \frac{1}{u\theta} h_\theta^*\left(\frac{1}{u} - (m-1)\theta\right), \quad u \in (0, \theta], \quad (2.9.8)$$

and we indeed have $U_\theta g_\theta^* = h_\theta^*$ since

$$U_\theta g_\theta^*(x) = \sum_{j \geq m} \frac{1+\theta x}{(x+j\theta)(x+(j+1)\theta)} g_\theta^*\left(\frac{1}{x+j\theta}\right)$$

$$= \sum_{j \geq m} \frac{1+\theta x}{\theta} \left\{ \frac{h_\theta^*(x+j\theta - m\theta)}{x+j\theta} - \frac{h_\theta^*(x+(j+1)\theta - m\theta)}{x+(j+1)\theta} \right\}$$

$$= \frac{1+\theta x}{\theta} \left(\frac{h_\theta^*(x)}{x+m\theta} - \lim_{j \to \infty} \frac{h_\theta^*(x+(j+1)\theta - m\theta)}{x+(j+1)\theta} \right) = h_\theta^*(x), \quad x \in \mathbb{R}_+. \quad (2.9.9)$$

In particular, for any fixed $b_\theta \in [0, \theta]$ we consider the function $h_{b_\theta}^* : \mathbb{R}_+ \to \mathbb{R}$ defined as

$$h_{b_\theta}(x) := \frac{1}{e_\theta x + b_\theta + 1}, \quad x \in \mathbb{R}_+, \quad (2.9.10)$$

where the coefficient e_θ will be specified later. By the above, the function $g_{b_\theta}^* : (0, \theta] \to \mathbb{R}$ defined as

$$g_{b_\theta}^*(x) = \left(\frac{1}{\theta x} + 1\right) h_{b_\theta}^*\left(\frac{1}{x} - m\theta\right) - \frac{1}{\theta x} h_{b_\theta}^*\left(\frac{1}{x} - (m-1)\theta\right)$$

$$= \left(\frac{1}{\theta x} + 1\right) \frac{1}{e_\theta\left(\frac{1}{x} - m\theta\right) + b_\theta + 1} - \frac{1}{\theta x} \frac{1}{e_\theta\left(\frac{1}{x} - (m-1)\theta\right) + b_\theta + 1}$$

$$= \frac{1}{\theta} \left(\frac{x\theta + 1}{e_\theta + x(-e_\theta m\theta + b_\theta + 1)} - \frac{1}{e_\theta + x(-e_\theta(m-1)\theta + b_\theta + 1)} \right)$$
(2.9.11)

for any $x \in (0, \theta]$ satisfies

$$U_\theta g_{b_\theta}^*(x) = h_{b_\theta}^*(x), \quad x \in [0, \theta]. \quad (2.9.12)$$

Setting

$$\varphi_{b_\theta}(x) = (g_{b_\theta}^*)'(x)$$
$$= \frac{1}{\theta}\left(\frac{e_\theta(m+1)\theta - b_\theta - 1}{(e_\theta + x(-e_\theta m\theta + b_\theta + 1))^2} - \frac{e_\theta(m-1)\theta - b_\theta - 1}{(e_\theta + x(-e_\theta(m-1)\theta + b_\theta + 1))^2}\right)$$
(2.9.13)

we have

$$U_\theta^* \varphi_{b_\theta}(x) = -(U_\theta g_{b_\theta}^*)'(x) = -(h_{b_\theta}^*)'(x) = \frac{e_\theta}{(e_\theta x + b_\theta + 1)^2}, \quad x \in [0, \theta]. \quad (2.9.14)$$

We choose b_θ by asking that $(\varphi_{b_\theta}/U_\theta^*\varphi_{b_\theta})(0) = (\varphi_{b_\theta}/U_\theta^*\varphi_{b_\theta})(\theta)$. Since

$$(\varphi_{b_\theta}/U_\theta^*\varphi_{b_\theta})(0) = \frac{2(b_\theta + 1)^2}{e_\theta^2} \quad (2.9.15)$$

and

$$(\varphi_{b_\theta}/U_\theta^*\varphi_{b_\theta})(\theta) = \frac{(b_\theta + 1 + e_\theta\theta)^2}{e_\theta\theta^3}\left(\frac{e_\theta(m+1)\theta - b_\theta - 1}{(b_\theta + 1)^2} - \frac{e_\theta(m-1)\theta - b_\theta - 1}{(b_\theta + 1 + e_\theta\theta)^2}\right), \quad (2.9.16)$$

this amounts to the equation

$$H_\theta(b_\theta) = 2\theta(b_\theta + 1)^4 - e_\theta^3\{(2m+1)(b_\theta + 1) + e_\theta(m+1)\theta\} = 0. \quad (2.9.17)$$

We choose the coefficient e_θ such that the equation $H_\theta(x) = 0$, $x \in [0, \theta]$, yields a unique solution $b_\theta \in [0, \theta]$. Asking that

$$H_\theta(0) < 0, \quad H_\theta(\theta) > 0, \quad \text{and} \quad \frac{dH_\theta}{db_\theta} > 0, \quad (2.9.18)$$

we may determine e_θ (see Appendix A1). For this unique acceptable solution $b_\theta \in [0, \theta]$ the function $\varphi_{b_\theta}/U_\theta^*\varphi_{b_\theta}$ attains its maximum equal to $2(b_\theta + 1)^2/e_\theta^2$ at $x = 0$ and $x = \theta$ and has a minimum $m(b_\theta) = (\varphi_{b_\theta}/U_\theta^*\varphi_{b_\theta})(x_{min}^\theta) > 1$ (see Appendix A2). It follows that for $\varphi_\theta = \varphi_{b_\theta}$ we have

$$\frac{e_\theta^2 \varphi_\theta}{2(b_\theta + 1)^2} \leq U_\theta^* \varphi_\theta \leq \frac{\varphi_\theta}{m(b_\theta)}, \quad (2.9.19)$$

2.9 A Near-Optimal Solution to the Gauss-Kuzmin-Lévy Problem

that is, $v_\theta \varphi_\theta \leq U_\theta^* \varphi_\theta \leq w_\theta \varphi_\theta$, where

$$v_\theta = \frac{e_\theta^2}{2(b_\theta + 1)^2} \quad \text{and} \quad w_\theta = \frac{1}{m(b_\theta)}. \tag{2.9.20}$$

□

Corollary 2.9.2 ([63]) *Let* $f_{\theta,0} \in C^1([0,\theta])$ *such that* $(f_{\theta,0})' > 0$. *Put* $\alpha_\theta^* = \min_{x \in [0,\theta]} \varphi_\theta(x)/(f_{\theta,0})'(x)$ *and* $\beta_\theta^* = \max_{x \in [0,\theta]} \varphi_\theta(x)/(f_{\theta,0})'(x)$. *Then*

$$\frac{\alpha_\theta^*}{\beta_\theta^*} v_\theta^n (f_{\theta,0})' \leq (U_\theta^*)^n (f_{\theta,0})' \leq \frac{\beta_\theta^*}{\alpha_\theta^*} w_\theta^n (f_{\theta,0})', \quad n \in \mathbb{N}_+. \tag{2.9.21}$$

Proof Since U_θ^* is a positive operator, we have

$$v_\theta^n \varphi_\theta \leq (U_\theta^*)^n \varphi_\theta \leq w_\theta^n \varphi_\theta, \quad n \in \mathbb{N}_+. \tag{2.9.22}$$

Noting that $\alpha_\theta^* (f_{\theta,0})' \leq \varphi_\theta \leq \beta_\theta^* (f_{\theta,0})'$, we can write

$$\frac{\alpha_\theta^*}{\beta_\theta^*} v_\theta^n (f_{\theta,0})' \leq \frac{v_\theta^n}{\beta_\theta^*} \varphi_\theta \leq \frac{1}{\beta_\theta^*} (U_\theta^*)^n \varphi_\theta \leq (U_\theta^*)^n (f_{\theta,0})'$$

$$\leq \frac{1}{\alpha_\theta^*} w_\theta^n \varphi_\theta \leq \frac{\beta_\theta^*}{\alpha_\theta^*} w_\theta^n (f_{\theta,0})', \quad n \in \mathbb{N}_+, \tag{2.9.23}$$

which shows that (2.9.21) holds. □

Theorem 2.9.3 (Near-Optimal Solution to Gauss-Kuzmin-Lévy Problem [63]) *Let* $f_{\theta,0} \in C^1([0,\theta])$ *such that* $(f_{\theta,0})' > 0$ *and let* μ *be a probability measure on* $\mathcal{B}_{[0,\theta]}$ *such that* $\mu \ll \lambda_\theta$. *For any* $n \in \mathbb{N}_+$ *and* $x \in [0,\theta]$ *we have*

$$(\log(1+\theta^2))^2 \frac{\alpha_\theta^*}{2\theta \beta_\theta^*} \min_{x \in [0,\theta]} (f_{\theta,0})'(x) v_\theta^n F_\theta(x)(\theta - F_\theta(x)) \leq |\mu(T_\theta^n < x) - F_\theta(x)|$$

$$\leq (\log(1+\theta^2))^2 \frac{(1+\theta^2)\beta_\theta^*}{2\theta \alpha_\theta^*} \max_{x \in [0,\theta]} (f_{\theta,0})'(x) w_\theta^n F_\theta(x)(\theta - F_\theta(x)) \tag{2.9.24}$$

where α_θ^*, β_θ^*, v_θ, *and* w_θ *are defined in Proposition 2.9.1 and Corollary 2.9.2, and*

$$F_\theta(x) = \frac{\log(1+\theta x)}{\log(1+\theta^2)}. \tag{2.9.25}$$

Proof For any $n \in \mathbb{N}$ and $x \in [0, \theta]$ set $d_n(F_\theta(x)) = \mu(T_\theta^n < x) - F_\theta(x)$. Then by (2.9.1) we have

$$d_n(F_\theta(x)) = \int_0^x \frac{U_\theta^n f_{\theta,0}(u)}{(1+\theta u)} \theta du - F_\theta(x). \tag{2.9.26}$$

Differentiating twice with respect to x yields

$$d_n'(F_\theta(x)) \frac{1}{\log(1+\theta^2)} \frac{\theta}{1+\theta x} = \frac{U_\theta^n f_{\theta,0}(x)}{1+\theta x} \theta - \frac{1}{\log(1+\theta^2)} \frac{\theta}{1+\theta x}, \tag{2.9.27}$$

$$(U_\theta^n f_{\theta,0}(x))' = \frac{\theta}{\left(\log(1+\theta^2)\right)^2} \frac{d_n''(F_\theta(x))}{1+\theta x}, \quad n \in \mathbb{N}, x \in [0, \theta]. \tag{2.9.28}$$

Hence by (2.9.3) we have

$$d_n''(F_\theta(x)) = (-1)^n \left(\log(1+\theta^2)\right)^2 \frac{1+\theta x}{\theta} (U_\theta^*)^n (f_{\theta,0})'(x), \tag{2.9.29}$$

for any $n \in \mathbb{N}$, $x \in [0, \theta]$. Since $d_n(0) = d_n(\theta) = 0$, a well-known interpolation formula yields

$$d_n(x) = -\frac{x(\theta - x)}{2} d_n''(\xi), \quad n \in \mathbb{N}, x \in [0, \theta], \tag{2.9.30}$$

for a suitable $\xi = \xi(n, x) \in [0, \theta]$. Therefore

$$\mu(T_\theta^n < x) - F_\theta(x)$$
$$= (-1)^{n+1} \left(\log(1+\theta^2)\right)^2 \frac{\theta \xi_\theta + 1}{\theta} (U_\theta^*)^n (f_\theta^0)'(\xi_\theta) \frac{F_\theta(x)(\theta - F_\theta(x))}{2}$$

for any $n \in \mathbb{N}$ and $x \in [0, \theta]$, and another suitable $\xi_\theta = \xi_\theta(n, x) \in [0, \theta]$. The result stated follows now from Corollary 2.9.2. $\square$

2.9.1 Appendix A1

To conclude, we use the values obtained in the Appendix A2 (see [63]).

Let us consider the case $m = 2$. The equation $H_\theta(x) = 0$, with $e_\theta = 1$, has as unique acceptable solution $b_\theta = 0.6445398$. For this value of b_θ the function $\varphi_{b_\theta}/U_\theta^* \varphi_{b_\theta}$ attains its maximum equal to 5.409022308 at $x = 0$ and $x = \theta$ and has a minimum $m(b_\theta) = (\varphi_{b_\theta}/U_\theta^* \varphi_{b_\theta})(0.297421) = 5.28441$. It follows that upper and lower bounds of the convergence rate are respectively $O(w_\theta^n)$ and $O(v_\theta^n)$ as $n \to \infty$, with $v_\theta > 0.184876294$ and $w_\theta < 0.189235884$.

2.9 A Near-Optimal Solution to the Gauss-Kuzmin-Lévy Problem

For $m=3$, the equation $H_\theta(x) = 0$, with $e_\theta = 0.67$, has as unique acceptable solution $b_\theta = 0.287897$. For this value of b_θ the function $\varphi_{b_\theta}/U_\theta^* \varphi_{b_\theta}$ attains its maximum equal to 7.389969626 at $x=0$ and $x=\theta$ and has a minimum $m(b_\theta) = (\varphi_{b_\theta}/U_\theta^* \varphi_{b_\theta})(0.256122) = 7.29924$. It follows that upper and lower bounds of the convergence rate are respectively $O(w_\theta^n)$ and $O(v_\theta^n)$ as $n \to \infty$, with $v_\theta > 0.135318553$ and $w_\theta < 0.137000564$.

Finally, let us consider the case $m=4$. The equation $H_\theta(x) = 0$, with $e_\theta = 0.5772$, has as unique acceptable solution $b_\theta = 0.249911$. For this value of b_θ the function $\varphi_{b_\theta}/U_\theta^* \varphi_{b_\theta}$ attains its maximum equal to 9.378546393 at $x=0$ and $x=\theta$ and has a minimum $m(b_\theta) = (\varphi_{b_\theta}/U_\theta^* \varphi_{a\theta})(0.228161) = 9.3072$. It follows that upper and lower bounds of the convergence rate are respectively $O(w_\theta^n)$ and $O(v_\theta^n)$ as $n \to \infty$, with $v_\theta > 0.106626331$ and $w_\theta < 0.107443699$.

2.9.2 Appendix A2

A2.1 Imposing conditions (2.9.18) and using a mathematical software, we obtain:

m	e_θ	b_θ	m	e_θ	b_θ
2	1	0.6445398	17	0.266721	0.121272
3	0.6704	0.287897	18	0.258692	0.117853
4	0.5772	0.249911	19	0.251324	0.114708
5	0.513167	0.223606	20	0.244533	0.111805
6	0.465794	0.204125	25	0.217106	0.100000
7	0.429017	0.188983	30	0.197052	0.0912886
8	0.399444	0.176778	35	0.181587	0.0845132
9	0.375022	0.166667	40	0.169204	0.0790571
10	0.354429	0.158113	45	0.159005	0.0745384
11	0.336772	0.150756	50	0.150419	0.0707134
12	0.321422	0.144338	60	0.136674	0.0645535
13	0.307923	0.138677	70	0.126065	0.0597579
14	0.295935	0.133633	80	0.117564	0.0559056
15	0.285198	0.129098	90	0.110555	0.0527013
16	0.275514	0.125003	100	0.104651	0.0499983

A2.2 Since

$$(\varphi_{b_\theta}/U_\theta^* \varphi_{b_\theta})'(x) = \frac{2}{\theta e_\theta}(e_\theta x + b_\theta + 1)\left[\frac{(A_\theta + \theta e_\theta)(e_\theta^2 + A_\theta(b_\theta + 1))}{(e_\theta - xA_\theta)^3} - \frac{(A_\theta - \theta e_\theta)(e_\theta^2 + (A_\theta - \theta e_\theta)(b_\theta + 1))}{(e_\theta - x(A_\theta - \theta e_\theta))^3}\right],$$

where $A_\theta := \theta e_\theta m - b_\theta - 1$, the equation $(\varphi_{b_\theta}/U_\theta^* \varphi_{b_\theta})'(x) = 0$ has a unique positive solution in $(0, \theta)$

$$x_{min}^\theta = \frac{e_\theta (B_\theta - C_\theta)}{(A_\theta - \theta e_\theta) B_\theta - A_\theta C_\theta}$$

with

$$B_\theta := \sqrt[3]{(A_\theta + \theta e_\theta)(e_\theta^2 + (b_\theta + 1) A_\theta)}, \quad C_\theta := \sqrt[3]{(A_\theta - \theta e_\theta)(e_\theta^2 + (b_\theta + 1)(A_\theta - \theta e_\theta))}.$$

Since in the particular cases studied for $m = 2$, $m = 3$, and $m = 4$, the minimum $m(b_\theta) = (\varphi_{b_\theta}/U_\theta^* \varphi_{b_\theta})(x_{min}^\theta)$ has the following values:

$m = 2$	$m(b_\theta) = 5.28441\ldots$
$m = 3$	$m(b_\theta) = 7.29924\ldots$
$m = 4$	$m(b_\theta) = 9.3072\ldots$

we may assume that $m(b_\theta) > 1$ for every $m \in \mathbb{N}_+$.

2.10 Gauss-Kuzmin Theorem Related to the Natural Extension

Let $\left([0,\theta]^2, \mathcal{B}_{[0,\theta]}^2, \overline{\gamma}_\theta, \overline{T}_\theta\right)$ be as in Sect. 2.4.1. In this section a Gauss-Kuzmin theorem for $\left([0,\theta]^2, \mathcal{B}_{[0,\theta]}^2, \overline{\gamma}_\theta, \overline{T}_\theta\right)$ is given. First we give a modified version of the Gauss-Kuzmin theorem for T_θ proved above. Then we show some important results used in the proof of the main theorem.

Theorem 2.10.1 ([65]) *Let* $([0, \theta], \mathcal{B}_{[0,\theta]}, \gamma_\theta, T_\theta)$ *as in Definition 2.1.1. There exists a constant* $0 < q_\theta < 1$ *such that for any* $A \in \mathcal{B}_{[0,\theta]}$ *we have*

$$\left| \lambda_\theta \left(T_\theta^{-n}(A) \right) - \gamma_\theta(A) \right| < C_\theta q_\theta^n \lambda_\theta(A), \tag{2.10.1}$$

where C_θ is a universal constant.

Proof In Proposition 2.5.1(iii) we show that $\mu\left(T_\theta^{-n}(A)\right) = \int_A U_\theta^n f_\theta(x) d\gamma_\theta(x)$, where μ is a probability measure on $\left([0, \theta], \mathcal{B}_{[0,\theta]}\right)$ absolutely continuous with respect to the Lebesgue measure λ_θ, and $f_\theta(x) := (\log(1 + \theta^2)) \frac{1+\theta x}{\theta^2} h_\theta(x)$ with $h_\theta := d\mu/d\lambda_\theta$ a.e. in $[0, \theta]$. In the special case $\mu = \lambda_\theta$ we obviously have

2.10 Gauss-Kuzmin Theorem Related to the Natural Extension

$$\lambda_\theta \left(T_\theta^{-n}(A)\right) = \frac{\theta}{\log\left(1+\theta^2\right)} \int_A \frac{U_\theta^n f_\theta(x)}{1+\theta x} dx \qquad (2.10.2)$$

with $f_\theta(x) := (\log(1+\theta^2))\frac{1+\theta x}{\theta^2}$, $x \in [0,\theta]$. Thus, from (2.5.10) we have that $U_\theta^\infty f_\theta = 1$. Therefore,

$$\gamma_\theta(A) = \frac{\theta}{\log\left(1+\theta^2\right)} \int_A \frac{U_\theta^\infty f_\theta}{1+\theta x} dx. \qquad (2.10.3)$$

In Sect. 2.6 we showed that there exist two positive constants $q_\theta < 1$ and K_θ such that

$$\left\|U_\theta^n f - U_\theta^\infty f\right\|_L \le K_\theta q_\theta^n \|f\|_L, \quad f \in L([0,\theta]), n \in \mathbb{N}_+.$$

Therefore

$$\left|\lambda_\theta \left(T_\theta^{-n}(A)\right) - \gamma_\theta(A)\right| \le \frac{\theta}{\log\left(1+\theta^2\right)} \int_A \frac{\left|U_\theta^n f_\theta(x) - U_\theta^\infty f_\theta\right|}{1+\theta x} dx$$

$$< K_\theta q_\theta^n \|f_\theta\|_L \frac{\theta}{\log\left(1+\theta^2\right)} \int_A \frac{1}{1+\theta x} dx = K_\theta q_\theta^n \|f_\theta\|_L \gamma_\theta(A),$$

and since

$$\gamma_\theta(A) \le \frac{\theta}{\log\left(1+\theta^2\right)} \lambda_\theta(A), \quad A \in \mathcal{B}_{[0,\theta]}$$

then the proof is complete. □

We now give the main theorem of this section. First, define $\Delta_{x,y} := [0,x] \times [0,y]$ for any $x, y \in [0,\theta]$, and the functions $(\overline{F}_{\theta,n})_{n\in\mathbb{N}_+}$ on $[0,\theta]^2$ by

$$\overline{F}_{\theta,n}(x,y) := \overline{\lambda}_\theta \left(\left\{(\xi,\zeta) \in [0,\theta]^2 : \left(\overline{T}_\theta\right)^n (\xi,\zeta) \in \Delta_{x,y}\right\}\right), \qquad (2.10.4)$$

where $\overline{\lambda}_\theta$ is the Lebesgue measure on $[0,\theta]^2$.

Theorem 2.10.2 (A Gauss-Kuzmin Theorem for $\overline{T}_\theta$ [65]) *For every $n \ge 2$ and $(x,y) \in [0,\theta]^2$ one has*

$$\overline{F}_{\theta,n}(x,y) = \frac{\log(1+xy)}{\log(1+\theta^2)} + \mathcal{O}\left((\overline{\alpha}_\theta)^n\right) \qquad (2.10.5)$$

with $0 < \overline{\alpha}_\theta < 1$.

Like in the one-dimensional case, we need the Gauss-Kuzmin-type equation associated with the functions $(\overline{F}_{\theta,n})_{n\in\mathbb{N}_+}$ defined in (2.10.4). Thus, for any $0 < y \leq \theta$, put $\ell_1 := b_1(y)$, where b_1 is as in (2.1.5). Then $(\overline{T}_\theta)^{n+1}(x, y) \in \Delta_{x,y}$ is equivalent to

$$(\overline{T}_\theta)^n \in \left\{\bigcup_{i\geq \ell_1+1}\left[\frac{1}{i\theta+x}, \frac{1}{i\theta}\right] \times [0,\theta]\right\} \cup \left\{\left[\frac{1}{\ell_1\theta+x}, \frac{1}{\ell_1\theta}\right] \times \left[\frac{1}{y}-\ell_1\theta, \theta\right]\right\}.$$

From this and (2.10.4) we get the Gauss-Kuzmin-type equation on $[0,\theta]^2$:

$$\overline{F}_{\theta,n+1}(x,y) = \sum_{i\geq \ell_1}\left\{\overline{F}_{\theta,n}\left(\frac{1}{i\theta},\theta\right) - \overline{F}_{\theta,n}\left(\frac{1}{i\theta+x},\theta\right)\right\}$$

$$-\left\{\overline{F}_{\theta,n}\left(\frac{1}{\ell_1\theta},\frac{1}{y}-\ell_1\theta\right) - \overline{F}_{\theta,n}\left(\frac{1}{\ell_1\theta+x},\frac{1}{y}-\ell_1\theta\right)\right\}. \quad (2.10.6)$$

A straightforward calculation shows that the density of the measure $\overline{\gamma}_\theta$ defined in (2.4.7) is an eigenfunction of (2.10.5), namely, if we put $\overline{F}_{\theta,n}(x,y) = \log(1+xy)$, $x, y \in [0,\theta]$, we obtain $\overline{F}_{\theta,n+1}(x,y) = \log(1+xy)$.

Lemma 2.10.3 ([65]) Let $n \in \mathbb{N}$, $n \geq 2$ and let $y \in [0,\theta] \cap \mathbb{Q}$ with $y = [\ell_1\theta,\ldots,\ell_d\theta]$, $\ell_1,\ldots,\ell_d \in \mathbb{N}_m$, $\ell_d \geq m+1$, where $d \leq \lfloor n/2 \rfloor$. Then for every $x, x^* \in [0,\theta)$ with $x^* < x$,

$$\left|\overline{F}_{\theta,n}(x,y) - \overline{F}_{\theta,n}(x^*,y) - \frac{1}{\log(1+\theta^2)}\log\left(\frac{1+xy}{1+x^*y}\right)\right|$$

$$< \frac{(1+\theta^2)^2}{\theta^2}\frac{C_\theta}{1-q_\theta}\overline{\lambda}_\theta(\Delta_{x,y}\setminus\Delta_{x^*,y})q_\theta^{n-d},$$

where $0 < q_\theta < 1$ and C_θ as given in Theorem 2.10.1.

Proof of Theorem 2.10.2 Let $(x,y) \in [0,\theta]^2$, $n \geq 2$. In view of Lemma 2.10.3 we assume that $y \notin \mathbb{Q}$. Put $d := \max\{\kappa \in \mathbb{N} : \kappa \text{ is even and } \kappa \leq \lfloor n/2 \rfloor + 2\}$. Using (2.2.2) we obtain

$$\left|y - \frac{p_d}{q_d}\right| \leq \frac{1}{q_d q_{d+1}} < \frac{1}{q_d^2}. \quad (2.10.7)$$

Furthermore, $q_n \geq \overline{q}_n$, $n \in \mathbb{N}_+$, where $\overline{q}_n$ satisfies the recurrence relation $\overline{q}_n = m\theta\overline{q}_{n-1} + \overline{q}_{n-2}$, $n \in \mathbb{N}_+$, with $\overline{q}_{-1} = 0$ and $\overline{q}_0 = 1$. It is easy to see that

2.10 Gauss-Kuzmin Theorem Related to the Natural Extension

$$\overline{q}_n = \frac{\theta}{\sqrt{1+4\theta^2}} \left\{ \left(\frac{1+\sqrt{1+4\theta^2}}{2\theta} \right)^{n+1} - \left(\frac{1-\sqrt{1+4\theta^2}}{2\theta} \right)^{n+1} \right\}. \qquad (2.10.8)$$

From (2.10.7) and (2.10.8) and the fact that d is even, we get

$$\left| y - \frac{p_d}{q_d} \right| < \frac{1}{(\overline{q}_d)^2} < \frac{1+4\theta^2}{\theta^2} \frac{1}{\left(\frac{1+\sqrt{1+4\theta^2}}{2\theta} \right)^n}. \qquad (2.10.9)$$

Now for each $B \in \mathcal{B}^2_{[0,\theta]}$ one has

$$\frac{1}{(1+\theta^2)} \frac{1}{\log(1+\theta^2)} \overline{\lambda}_\theta(B) \leq \overline{\gamma}_\theta(B) \leq \frac{1}{\log(1+\theta^2)} \overline{\lambda}_\theta(B). \qquad (2.10.10)$$

Since $\Delta_{x, p_d/q_d} \subset \Delta_{x,y}$ and $\overline{F}_{\theta,n}(x,y) = \overline{\lambda}_\theta \left((\overline{T}_\theta)^{-n} (\Delta_{x,y}) \right)$, from (2.10.10), (2.10.9) and the fact that $\overline{T}_\theta$ is $\overline{\gamma}_\theta$-invariant, we find that

$$\overline{F}_{\theta,n}(x,y) - \overline{F}_{\theta,n}\left(x, \frac{p_d}{q_d}\right) = \overline{\lambda}_\theta \left((\overline{T}_\theta)^{-n} (\Delta_{x,y}) \setminus (\overline{T}_\theta)^{-n} (\Delta_{x,p_d/q_d}) \right)$$
$$\leq (1+\theta^2) \log(1+\theta^2) \overline{\gamma}_\theta \left((\overline{T}_\theta)^{-n} (\Delta_{x,y}) \setminus (\overline{T}_\theta)^{-n} (\Delta_{x,p_d/q_d}) \right)$$
$$= (1+\theta^2) \log(1+\theta^2) \overline{\gamma}_\theta \left((\overline{T}_\theta)^{-n} (\Delta_{x,y} \setminus \Delta_{x,p_d/q_d}) \right)$$
$$\leq (1+\theta^2) \log(1+\theta^2) \frac{1}{\log(1+\theta^2)} \overline{\lambda}_\theta \left([0,x] \times \left[\frac{p_d}{q_d}, y \right] \right)$$
$$= (1+\theta^2) \frac{x}{\theta} \frac{1}{\theta} \left| y - \frac{p_d}{q_d} \right| < \frac{1+\theta^2}{\theta^2} \frac{1+4\theta^2}{\theta^2} \frac{x}{\left(\frac{1+\sqrt{1+4\theta^2}}{2\theta} \right)^n}. \qquad (2.10.11)$$

Since for every fixed $x \in [0,\theta]$ the function $y \mapsto \log(1+xy)$ is differentiable on $[0,\theta]$, by the *Mean Value Theorem* we have

$$\left| \log(1+xy) - \log\left(1 + x\frac{p_d}{q_d}\right) \right| = \left| y - \frac{p_d}{q_d} \right| \cdot \left| \frac{x}{1+x\xi} \right|$$
$$< \frac{1+4\theta^2}{\theta^2} \frac{x}{\left(\frac{1+\sqrt{1+4\theta^2}}{2\theta} \right)^n}, \qquad (2.10.12)$$

where $p_d/q_d \leq \xi \leq y$.

Finally, from Lemma 2.10.3, (2.10.11), and (2.10.12), and since $\overline{F}_{\theta,n}(0, p_d/q_d) = 0$, we have

$$\left|\overline{F}_{\theta,n}(x, y) - \frac{\log(1+xy)}{\log(1+\theta^2)}\right| \leq \left|\overline{F}_{\theta,n}(x, y) - \overline{F}_{\theta,n}\left(x, \frac{p_d}{q_d}\right)\right|$$

$$+ \left|\overline{F}_{\theta,n}\left(x, \frac{p_d}{q_d}\right) - \overline{F}_{\theta,n}\left(0, \frac{p_d}{q_d}\right)\right.$$

$$\left. - \frac{1}{\log(1+\theta^2)} \log\left(1 + x\frac{p_d}{q_d}\right)\right|$$

$$+ \frac{1}{\log(1+\theta^2)} \left|\log(1+xy) - \log\left(1 + x\frac{p_d}{q_d}\right)\right|$$

$$\leq \frac{1+\theta^2}{\theta^2} \frac{1+4\theta^2}{\theta^2} \frac{x}{\left(\frac{1+\sqrt{1+4\theta^2}}{2\theta}\right)^n}$$

$$+ \frac{(1+\theta^2)^2}{\theta^2} \frac{C_\theta}{1-q_\theta} \frac{x}{\theta^2} \frac{p_d}{q_d} q_\theta^{n-d}$$

$$+ \frac{1}{\log(1+\theta^2)} \frac{1+4\theta^2}{\theta^2} \frac{x}{\left(\frac{1+\sqrt{1+4\theta^2}}{2\theta}\right)^n}$$

$$\leq \frac{1+4\theta^2}{\theta} \left(\frac{1+\theta^2}{\theta^2} + \frac{1}{\log(1+\theta^2)}\right) \frac{1}{\left(\frac{1+\sqrt{1+4\theta^2}}{2\theta}\right)^n}$$

$$+ \frac{(1+\theta^2)^2}{\theta^2} \frac{C_\theta}{1-q_\theta} q_\theta^{n-d}.$$

Since $d \leq n/2 + 2$, we have

$$\left|\overline{F}_{\theta,n}(x, y) - \frac{\log(1+xy)}{\log(1+\theta^2)}\right| \leq \frac{1+4\theta^2}{\theta} \left(\frac{1+\theta^2}{\theta^2} + \frac{1}{\log(1+\theta^2)}\right) \frac{1}{\left(\frac{1+\sqrt{1+4\theta^2}}{2\theta}\right)^n}$$

$$+ \frac{(1+\theta^2)^2}{\theta^2} \frac{C_\theta (\sqrt{q_\theta})^n}{(1-q_\theta) q_\theta^2} \leq \mathcal{K}_\theta \overline{\alpha}_\theta^n,$$

where

$$K_\theta := \frac{1+4\theta^2}{\theta}\left(\frac{1+\theta^2}{\theta^2} + \frac{1}{\log(1+\theta^2)}\right) + \frac{(1+\theta^2)^2}{\theta^2}\frac{C_\theta}{(1-q_\theta)q_\theta^2}$$

and

$$\overline{\alpha}_\theta := \max\left(\sqrt{q_\theta}, \frac{2\theta}{1+\sqrt{1+4\theta^2}}\right) < 1.$$

$\square$

2.11 Two Asymptotic Distributions

In this section we shall estimate the error term

$$e_{\theta,n}(\gamma_{\theta,b}; x, y) = \gamma_{\theta,b}\left(T_\theta^n \in [0,x], s_{\theta,n}^b \in [0,y]\right) - \frac{\log(1+xy)}{\log(1+\theta^2)}$$

for any $b \in [0, \theta]$, $x, y \in [0, \theta]$, and $n \in \mathbb{N}_+$, where $(\gamma_{\theta,b})_b$ is as in (2.4.14). In the main result of this section, Theorem 2.11.4, we shall derive lower and upper bounds (not depending on $b \in [0, \theta]$) of the supremum

$$\sup_{x,y \in [0,\theta]} |e_{\theta,n}(\gamma_{\theta,b}; x, y)|, \quad b \in [0, \theta], \tag{2.11.1}$$

which provide a more refined estimate of the convergence rate involved. First, we obtain a lower bound for the following approximation error.

Theorem 2.11.1 ([65]) *For any $b \in [0, \theta]$ and $n \in \mathbb{N}_+$ we have*

$$\frac{1}{2}P_{\theta,m(n)}(\theta) \le \sup_{y \in [0,\theta]} \left|\gamma_{\theta,b}\left(s_{\theta,n}^b \le y\right) - \gamma_\theta\left([0,y]\right)\right|$$

with $P_{\theta,m(n)}(\theta) = P_{\theta,i_1\ldots i_n}(\theta)$ as in (2.5.14), where we write $m(n)$ for $(i_1, \ldots, i_n)$ with $i_1 = \ldots = i_n = m$, $m \in \mathbb{N}_+$.

Proof The continuity of the function $y \to \gamma_\theta([0, y])$, $y \in [0, \theta]$ and the equation

$$\lim_{h \searrow 0} \gamma_{\theta,b}\left(s_{\theta,n}^b \le y - h\right) = \gamma_{\theta,b}\left(s_{\theta,n}^b \le y\right)$$

imply that

$$\sup_{y \in [0,\theta]} \left| \gamma_{\theta,b} \left(s_{\theta,n}^b \leq y \right) - \gamma_\theta \left([0, y] \right) \right| = \sup_{y \in [0,\theta]} \left| \gamma_{\theta,b} \left(s_{\theta,n}^b < y \right) - \gamma_\theta \left([0, y] \right) \right|$$

for any $b \in [0, \theta]$ and $n \in \mathbb{N}_+$. For any $s \in [0, \theta]$ we then have

$$\gamma_{\theta,b}(s_{\theta,n}^b = s) = \gamma_{\theta,b} \left(s_{\theta,n}^b \leq s \right) - \gamma_\theta \left([0, s] \right) - \left(\gamma_{\theta,b} \left(s_{\theta,n}^b < s \right) - \gamma_\theta \left([0, s] \right) \right)$$

$$\leq \sup_{y \in [0,\theta]} \left| \gamma_{\theta,b} \left(s_{\theta,n}^b \leq y \right) - \gamma_\theta \left([0, y] \right) \right|$$

$$+ \sup_{y \in [0,\theta]} \left| \gamma_{\theta,b} \left(s_{\theta,n}^b < y \right) - \gamma_\theta \left([0, y] \right) \right|$$

$$= 2 \sup_{y \in [0,\theta]} \left| \gamma_{\theta,b} \left(s_{\theta,n}^b \leq y \right) - \gamma_\theta \left([0, y] \right) \right|.$$

Hence

$$\sup_{y \in [0,\theta]} \left| \gamma_{\theta,b} \left(s_{\theta,n}^b \leq y \right) - \gamma_\theta \left([0, y] \right) \right| \geq \frac{1}{2} \sup_{s \in [0,\theta]} \gamma_{\theta,b} \left(s_{\theta,n}^b = s \right),$$

for any $b \in [0, \theta]$ and $n \in \mathbb{N}_+$. Next, using (2.5.14) we have

$$\gamma_{\theta,b} \left(s_{\theta,n}^b = [i_n \theta, \ldots, i_2 \theta, i_1 \theta + b] \right) = P_{\theta, i_1 \ldots i_n}(b), n \geq 2,$$

$$\gamma_{\theta,b} \left(s_{\theta,1}^b = \frac{1}{i_1 \theta + b} \right) = P_{\theta, i_1}(b)$$

for any $b \in [0, \theta]$ and $i_1, \ldots, i_n \in \mathbb{N}_m$. By (2.5.15) we have

$$\sup_{s \in [0,\theta]} \gamma_{\theta,b}(s_{\theta,n}^b = s) = P_{\theta, m(n)}(b), \quad b \in [0, \theta].$$

By the same equation we have

$$P_{\theta, m(n)}(b) = \frac{1 + b\theta}{q_{n-1}(\underbrace{m, \ldots, m}_{(n-1) \text{ times}})(b + m\theta) + p_{n-1}(\underbrace{m, \ldots, m}_{(n-1) \text{ times}})}$$

$$\times \frac{1}{q_n(\underbrace{m, \ldots, m, m}_{n \text{ times}})(b + m\theta) + p_n(\underbrace{m, \ldots, m, m}_{n \text{ times}})}.$$

2.11 Two Asymptotic Distributions

It is easy to see that $P_{\theta,m(n)}(\cdot)$ is a decreasing function. Therefore

$$\sup_{s\in[0,\theta]} \gamma_{\theta,b}\left(s_{\theta,n}^b = s\right) \geq P_{\theta,m(n)}(\theta), \quad n \in \mathbb{N}_+$$

for any $b \in [0, \theta]$. $\square$

Theorem 2.11.2 ([65] The Lower Bound)
For any $b \in [0, \theta]$ and $n \in \mathbb{N}_+$ we have

$$\frac{1}{2} P_{\theta,m(n)}(\theta) \leq \sup_{x,y\in[0,\theta]} \left|\gamma_{\theta,b}\left(T_\theta^n \in [0,x], s_{\theta,n}^b \in [0,y]\right) - \frac{\log(1+xy)}{\log(1+\theta^2)}\right|.$$

Proof For any $b \in [0, \theta]$ and $n \in \mathbb{N}_+$, by Theorem 2.11.1 we have

$$\sup_{x,y\in[0,\theta]} \left|\gamma_{\theta,b}\left(T_\theta^n \in [0,x], s_{\theta,n}^b \in [0,y]\right) - \frac{\log(1+xy)}{\log(1+\theta^2)}\right|$$

$$\geq \sup_{y\in[0,\theta]} \left|\gamma_{\theta,b}\left(T_\theta^n \in [0,\theta], s_{\theta,n}^b \in [0,y]\right) - \frac{\log(1+\theta y)}{\log(1+\theta^2)}\right|$$

$$= \sup_{y\in[0,\theta]} \left|\gamma_{\theta,b}\left(s_{\theta,n}^b \in [0,y]\right) - \gamma_\theta([0,y])\right| \geq \frac{1}{2} P_{\theta,m(n)}(\theta).$$

$\square$

In what follows we use the characteristic properties of the transition operator associated with the RSCC underlying θ-expansions. By restricting this operator to the Banach space of functions of bounded variation on $[0, \theta]$, we derive an explicit upper bound for the supremum (2.11.1).

Theorem 2.11.3 ([65] The Upper Bound) For any $b \in [0, \theta]$ and $n \in \mathbb{N}$ we have

$$\sup_{x,y\in[0,\theta]} \left|\gamma_{\theta,b}\left(T_\theta^n \in [0,x], s_{\theta,n}^b \in [0,y]\right) - \frac{\log(1+xy)}{\log(1+\theta^2)}\right| \leq \frac{1}{(m+1)^n}.$$

Proof Let $F_{\theta,n}^b(y) = \gamma_{\theta,b}(s_{\theta,n}^b \leq y)$ and $G_{\theta,n}^b(y) = F_{\theta,n}^b(y) - \gamma_\theta([0,y])$, $b, y \in [0, \theta]$, $n \in \mathbb{N}$. As we have noted U_θ is the transition operator of the Markov chain $\left(s_{\theta,n}^b\right)_{n\in\mathbb{N}}$ on $\left([0,\theta], \mathcal{B}_{[0,\theta]}, \gamma_{\theta,b}\right)$ for any $b \in [0, \theta]$. For any $y \in [0, \theta]$ consider the function $f_{\theta,y}$

defined on $[0, \theta]$ as

$$f_{\theta,y}(b) := \begin{cases} 1 \text{ if } 0 \leq b \leq y, \\ 0 \text{ if } y < b \leq \theta. \end{cases}$$

Hence $U_\theta^n f_{\theta,y}(b) = E_{\gamma_{\theta,b}}\left(f_{\theta,y}(s_{\theta,n}^b)\big| s_{\theta,0}^b = b\right) = \gamma_{\theta,b}(s_{\theta,n}^b \leq y)$ for all $b, y \in [0, \theta]$, $n \in \mathbb{N}$. As $U_\theta^\infty f_{\theta,y} = \int_0^\theta f_{\theta,y}(b)\gamma_\theta(db) = \gamma_\theta([0, y])$, $y \in [0, \theta]$. It follows from Proposition 2.5.5 that

$$|G_{\theta,n}^b(y)| = \left|\gamma_{\theta,b}(s_{\theta,n}^b \leq y) - \gamma_\theta([0, y])\right| = \left|U_\theta^n f_{\theta,y}(b) - U_\theta^\infty f_{\theta,y}\right|$$

$$\leq \frac{1}{(m+1)^n} \text{var } f_{\theta,y} = \frac{1}{(m+1)^n} \quad (2.11.2)$$

for all $b, y \in [0, \theta]$, $n \in \mathbb{N}$. By (2.4.15), for all $b, x, y \in [0, \theta]$ and $n \in \mathbb{N}$ we have

$$\gamma_{\theta,b}\left(T_\theta^n \in [0, x], s_{\theta,n}^b \in [0, y]\right) = \int_0^y \gamma_{\theta,b}\left(T_\theta^n \in [0, x]\big| s_{\theta,n}^b = z\right) dF_{\theta,n}^b(z)$$

$$= \int_0^y \frac{(1+\theta z)x}{(1+xz)\theta} dF_{\theta,n}^b(z) = \int_0^y \frac{(1+\theta z)x}{(1+xz)\theta} \gamma_\theta(dz) + \int_0^y \frac{(1+\theta z)x}{(1+xz)\theta} dG_{\theta,n}^b(z)$$

$$= \frac{\log(1+xy)}{\log(1+\theta^2)} + \frac{(1+\theta z)x}{(1+xz)\theta} G_{\theta,n}^b(z)\Big|_0^y - \int_0^y \frac{x(\theta-x)}{(1+xz)^2\theta} G_{\theta,n}^b(z) dz.$$

Hence, by (2.11.2)

$$\left|\gamma_{\theta,b}\left(T_\theta^n \in [0, x], s_{\theta,n}^b \in [0, y]\right) - \frac{\log(1+xy)}{\log(1+\theta^2)}\right|$$

$$\leq \frac{1}{(m+1)^n}\left(\frac{(1+\theta y)x}{(1+xy)\theta} - \frac{(\theta-x)xy}{(1+xy)\theta}\right)$$

$$= \frac{x}{(m+1)^n \theta} \leq \frac{1}{(m+1)^n}$$

for all $b, x, y \in [0, \theta]$ and $n \in \mathbb{N}$. $\square$

Combining Theorem 2.11.2 with Theorem 2.11.3, we obtain Theorem 2.11.4.

Theorem 2.11.4 ([65]) *For any $b \in [0, \theta]$ and $n \in \mathbb{N}_+$ we have*

$$\frac{1}{2} P_{\theta,m(n)}(\theta) \leq \sup_{x,y \in [0,\theta]} \left|\gamma_{\theta,b}\left(T_\theta^n \in [0, x], s_{\theta,n}^b \in [0, y]\right) - \frac{\log(1+xy)}{\log(1+\theta^2)}\right| \leq \frac{1}{(m+1)^n}.$$

2.11 Two Asymptotic Distributions

Remark 2.11.5 ([65]) Note that

$$P_{\theta,m(n)}(\theta) = \frac{m+1}{\overline{q}_{n+1}\overline{q}_{n+2}}, \quad n \in \mathbb{N}_+,$$

where $\overline{q}_n$ has the expression in (2.10.8). It should be noted that Theorem 2.11.4 in connection with the limits

$$\lim_{n \to \infty} \left(\frac{1}{2} P_{\theta,m(n)}(\theta)\right)^{1/n} = \frac{2\theta^2}{1 + 2\theta^2 + \sqrt{1+4\theta^2}},$$

$$\lim_{n \to \infty} \left(\frac{1}{(m+1)^n}\right)^{1/n} = \frac{1}{m+1},$$

leads to an estimate of the order of magnitude of the supremum (2.11.1). Actually, Theorem 2.11.4 implies that the convergence rate is $\mathcal{O}(\eta_\theta^n)$, with

$$\frac{2\theta^2}{1 + 2\theta^2 + \sqrt{1+4\theta^2}} \leq \eta_\theta \leq \frac{1}{m+1}.$$

For example, we have (Table 2.2):

Also, the graph bellow suggests that for very large values of m the lower and upper bounds are very close (Fig. 2.1).

Table 2.2 Lower and upper bounds of the convergence rate for some $m \geq 1$

$m = 1$	$g^2 = 0.381966 \leq \eta_\theta \leq 0.50000$
$m = 2$	$0.267949192 \leq \eta_\theta \leq 0.33333\ldots$
$m = 3$	$0.208711948 \leq \eta_\theta \leq 0.25000\ldots$
$m = 10$	$0.083920216 \leq \eta_\theta \leq 0.090909\ldots$
$m = 100$	$0.009804864 \leq \eta_\theta \leq 0.00990099$
$m = 1000$	$0.000998004 \leq \eta_\theta \leq 0.000999$
$m = 10000$	$0.00009998 \leq \eta_\theta \leq 0.00009999$

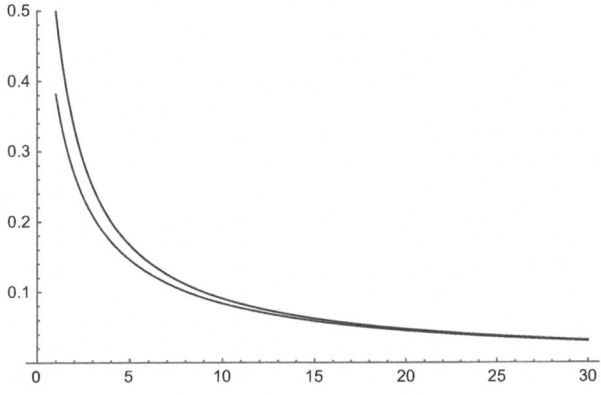

Fig. 2.1 Graphs of lower and upper bounds

The graphs for the lower and upper bounds were obtained from the functions

$$f_{lb} : \mathbb{N}_+ \to \mathbb{R}, \quad f_{lb}(m) = \frac{2}{m+2+\sqrt{m(m+4)}}$$

and, respectively,

$$f_{ub} : \mathbb{N}_+ \to \mathbb{R}, \quad f_{ub}(m) = \frac{1}{m+1}.$$

N-Continued Fractions 3

In this chapter we study a generalization of the regular continued fractions given by E.B. Burger et al. in 2008 [7]. We consider a family $\{T_N : N \geq 1\}$ of interval maps as generalizations of the Gauss transformation. For the continued fraction expansion arising from T_N, we solve its Gauss-Kuzmin-type problem by applying the theory of random systems with complete connections by M. Iosifescu. Then we give a two-dimensional Gauss-Kuzmin theorem for N-continued fraction expansions. More precisely, we obtain a Gauss-Kuzmin theorem related to the natural extension of the measure-theoretical dynamical system associated to this expansion.

3.1 Preliminary Considerations

We give here the basic metric properties of N-continued fraction expansions.

Fix an integer $N \geq 1$. In [7], Burger et al. proved that any irrational $0 < x < 1$ can be written in the form

$$x = \cfrac{N}{\varepsilon_1 + \cfrac{N}{\varepsilon_2 + \cfrac{N}{\varepsilon_3 + \cdots}}} =: [\varepsilon_1, \varepsilon_2, \varepsilon_3, \ldots]_N, \qquad (3.1.1)$$

where ε_n's are nonnegative integers. We will call (3.1.1) the *N-continued fraction expansion of x*. Here and below we omit the index N whenever it is clear from context.

This continued fraction is treated as the following dynamical system.

Definition 3.1.1 Fix an integer $N \geq 1$:

(i) The measure-theoretical dynamical system $(I, \mathcal{B}_I, T_N)$ is defined as follows: $I := [0, 1]$, $\mathcal{B}_I$ denotes the σ-algebra of all Borel subsets of I, and T_N is the transformation

$$T_N : I \to I; \quad T_N(x) := \begin{cases} \dfrac{N}{x} - \left\lfloor \dfrac{N}{x} \right\rfloor & \text{if } x \neq 0, \\ 0 & \text{if } x = 0. \end{cases} \tag{3.1.2}$$

(ii) In addition to (i), we write $(I, \mathcal{B}_I, G_N, T_N)$ as $(I, \mathcal{B}_I, T_N)$ with the following probability measure G_N on $(I, \mathcal{B}_I)$:

$$G_N(A) := \frac{1}{\log \frac{N+1}{N}} \int_A \frac{dx}{x + N}, \quad A \in \mathcal{B}_I. \tag{3.1.3}$$

Define the *quantized index map* $\eta : I \to \mathbb{N} := \mathbb{N}_+ \cup \{0\}$ by

$$\eta(x) := \begin{cases} \left\lfloor \dfrac{N}{x} \right\rfloor & \text{if } x \neq 0, \\ \infty & \text{if } x = 0. \end{cases} \tag{3.1.4}$$

By using T_N and η, the sequence $\{\varepsilon_n\}$ in (3.1.1) is obtained as follows:

$$\varepsilon_n = \varepsilon_n(x) := \eta\left(T_N^{n-1}(x)\right), \quad n \geq 1 \tag{3.1.5}$$

with $T_N^0(x) = x$. Putting $\mathbb{N}_N := \{N, N+1, \ldots\}$, $N \geq 1$, the incomplete quotients ε_n, $n \in \mathbb{N}_+$, take positive integer values in $\mathbb{N}_N$.

In this way, T_N gives the algorithm of N-continued fraction expansion which is an obvious generalization of the regular continued fraction.

Proposition 3.1.2 *Let $(I, \mathcal{B}_I, G_N, T_N)$ be as in Definition 3.1.1(ii):*

(i) $(I, \mathcal{B}_I, G_N, T_N)$ *is ergodic.*
(ii) *The measure G_N is invariant under T_N.*

Proof See [16] and Sect. 3.4. □

In [75], Van der Wekken showed the convergence of the expansion. For $x \in I \setminus \mathbb{Q} =: \Omega$, define the *n-th order convergent* $[\varepsilon_1, \varepsilon_2, \ldots, \varepsilon_n]_N$ of x by truncating the expansion on the

3.1 Preliminary Considerations

right-hand side of (3.1.1), that is,

$$[\varepsilon_1, \varepsilon_2, \ldots, \varepsilon_n]_N \to x, \quad n \to \infty. \tag{3.1.6}$$

To this end, for $n \in \mathbb{N}_+$, define integer-valued functions $p_{N,n}(x) = p_n(x)$ and $q_{N,n}(x) = q_n(x)$ by

$$p_n(x) := \varepsilon_n p_{n-1} + N p_{n-2}, \quad n \geq 2 \tag{3.1.7}$$

$$q_n(x) := \varepsilon_n q_{n-1} + N q_{n-2}, \quad n \geq 1 \tag{3.1.8}$$

with $p_0(x) = 0$, $q_0(x) = 1$, $p_{-1}(x) = 1$, $q_{-1}(x) = 0$, $p_1(x) = N$, $q_1(x) = \varepsilon_1$. By induction, we have

$$p_{n-1}(x)q_n(x) - p_n(x)q_{n-1}(x) = (-N)^n, \quad n \in \mathbb{N}. \tag{3.1.9}$$

By using (3.1.7) and (3.1.8), we can verify that

$$x = \frac{p_n(x) + T_N^n(x) p_{n-1}(x)}{q_n(x) + T_N^n(x) q_{n-1}(x)}, \quad n \geq 1. \tag{3.1.10}$$

By taking $T_N^n(x) = 0$ in (3.1.10), we obtain $[\varepsilon_1, \varepsilon_2, \ldots, \varepsilon_n]_N = p_n(x)/q_n(x)$. From this and by using (3.1.9) and (3.1.10), we obtain

$$\left| x - \frac{p_n(x)}{q_n(x)} \right| = \frac{N^n \cdot T_N^n(x)}{q_n(x) \left(T_N^n(x) q_{n-1}(x) + q_n(x) \right)}, \quad n \geq 1. \tag{3.1.11}$$

Now, since $T_N^n(x) < 1$ and $\left| T_N^n(x) \frac{q_{n-1}(x)}{q_n(x)} + 1 \right| \geq 1$, we have

$$\left| x - \frac{p_n(x)}{q_n(x)} \right| < \frac{N^n}{q_n^2(x)}, \quad n \geq 1. \tag{3.1.12}$$

In order to prove (3.1.6), it is sufficient to show the following inequality:

$$\left| x - \frac{p_n(x)}{q_n(x)} \right| \leq \frac{1}{N^n}, \quad n \geq 1. \tag{3.1.13}$$

From (3.1.8), we have that $q_n(x) > N q_{n-1}(x)$, and because $q_0 = 1$ we have $q_n(x) > N^n$. Finally, (3.1.13) follows from (3.1.12).

For $x \in \Omega$, let ε_n be as in (3.1.5). For any $n \in \mathbb{N}_+$ and $i^{(n)} = (i_1, \ldots, i_n) \in (\mathbb{N}_N)^n$, define the *fundamental interval associated with* $i^{(n)}$ by

$$I_N\left(i^{(n)}\right) = \{x \in \Omega : \varepsilon_k(x) = i_k \text{ for } k = 1, \ldots, n\}, \tag{3.1.14}$$

where we write $I_N\left(i^{(0)}\right) = \Omega$. Remark that $I_N\left(i^{(n)}\right)$ is not connected by definition. For example, we have

$$I_N(i) = \{x \in \Omega : \varepsilon_1 = i\} = \Omega \cap \left(\frac{N}{i+1}, \frac{N}{i}\right) \quad \text{for any } i \in \mathbb{N}_N. \tag{3.1.15}$$

Lemma 3.1.3 *Let λ denote the Lebesgue measure. Then*

$$\lambda\left(I_N\left(i^{(n)}\right)\right) = \frac{N^n}{q_n(x)(q_n(x) + q_{n-1}(x))}, \tag{3.1.16}$$

where $\{q_n\}$ is as in (3.1.8).

Proof From the definition of T_N and (3.1.10), we have

$$I_N\left(i^{(n)}\right) = \Omega \cap \left(u_N\left(i^{(n)}\right), v_N\left(i^{(n)}\right)\right), \tag{3.1.17}$$

where both $u_N\left(i^{(n)}\right)$ and $v_N\left(i^{(n)}\right)$ are rational numbers defined as

$$u_N\left(i^{(n)}\right) := \begin{cases} \dfrac{p_n(x) + p_{n-1}(x)}{q_n(x) + q_{n-1}(x)} & \text{if } n \text{ is odd,} \\ \dfrac{p_n(x)}{q_n(x)} & \text{if } n \text{ is even,} \end{cases} \tag{3.1.18}$$

and

$$v_N\left(i^{(n)}\right) := \begin{cases} \dfrac{p_n(x)}{q_n(x)} & \text{if } n \text{ is odd,} \\ \dfrac{p_n(x) + p_{n-1}(x)}{q_n(x) + q_{n-1}(x)} & \text{if } n \text{ is even.} \end{cases} \tag{3.1.19}$$

By using (3.1.9), we have (3.1.16). $\square$

We now give a Legendre-type result for N-continued fraction expansions. For $x \in \Omega$, we define the *approximation coefficient* $\Theta_N(x, n)$ by

$$\Theta_N(x, n) := \frac{q_n^2}{N^n}\left|x - \frac{p_n}{q_n}\right|, \quad n \geq 1, \tag{3.1.20}$$

where p_n/q_n is the n-th continued fraction convergent of x in (3.1.1).

3.1 Preliminary Considerations

Proposition 3.1.4 *For $x \in \Omega$ and an irreducible fraction $0 < p/q < 1$, assume that p/q is written as follows:*

$$\frac{p}{q} = [i_1, \ldots, i_n]_N, \tag{3.1.21}$$

where $[i_1, \ldots, i_n]_N$ is as in (3.1.1), and the length $n \in \mathbb{N}_+$ of N-continued fraction expansion of p/q is chosen in such a way that it is even if $p/q < x$ and odd otherwise. Then

$$\Theta_N(x, n) < \frac{q}{q + q_{n-1}} \quad \text{if and only if} \quad \frac{p}{q} \text{ is the } n-\text{th convergent of } x, \tag{3.1.22}$$

where $\Theta_N(x, n)$ is as in (3.1.20) and the positive integer q_{n-1} is defined as the denominator of the irreducible fraction representation of the rational number $[i_1, \ldots, i_{n-1}]_N$ with $q_0 = 1$ for the sequence $i_1, \ldots, i_n$.

Proof Fix $x \in \Omega$ and $n \geq 1$. Let $\Theta := \Theta_N(x, n)$.
($\Leftarrow$) Assume that p/q is the n-th convergent of x. By (3.1.11) and the definition of Θ, we have

$$\Theta = \frac{q^2}{N^n} \left| x - \frac{p}{q} \right| = \frac{T_N^n(x) q}{q + T_N^n(x) q_{n-1}(x)} \leq \frac{q}{q + q_{n-1}}, \tag{3.1.23}$$

where we use $q_{n-1} = q_{n-1}(x)$.
($\Rightarrow$) Conversely,

$$\text{if } \Theta < \frac{q}{q + q_{n-1}}, \quad \text{then} \quad q \left| x - \frac{p}{q} \right| < \frac{N^n}{q + q_{n-1}}. \tag{3.1.24}$$

If n is even, then $x > p/q$, and we have

$$x - \frac{p}{q} < \frac{N^n}{q(q + q_{n-1})}. \tag{3.1.25}$$

From these,

$$\frac{p}{q} < x < \frac{p}{q} + \frac{N^n}{q(q + q_{n-1})} = \frac{p + p_{n-1}}{q + q_{n-1}}, \tag{3.1.26}$$

where p_{n-1} is defined as $p_{n-1}/q_{n-1} = [i_1, \ldots, i_{n-1}]_N$. Hence $x \in I_N\left(i^{(n)}\right)$, i.e., $p/q = [i_1, \ldots, i_n]_N$ is a convergent of x. The case when n is odd is treated similarly. □

3.2 Basic Metric Properties

We derive the so-called Brodén-Borel-Lévy formula [28, 29] for N-continued fraction expansion. For $x \in I$, let ε_n and q_n be as in (3.1.5) and (3.1.8), respectively. We define $\left(s_n^N\right)_{n \geq 0}$ by

$$s_0^N := 0, \quad s_n^N := N \frac{q_{n-1}}{q_n}, \quad n \geq 1. \tag{3.2.1}$$

From (3.1.8), $s_n^N = N/(\varepsilon_n + s_{n-1}^N)$ for $n \geq 1$. Hence

$$s_n^N = \cfrac{N}{\varepsilon_n + \cfrac{N}{\varepsilon_{n-1} + \cdots + \cfrac{N}{\varepsilon_1}}} = [\varepsilon_n, \varepsilon_{n-1}, \ldots, \varepsilon_2, \varepsilon_1]_N, \tag{3.2.2}$$

for $n \geq 1$.

Proposition 3.2.1 (Brodén-Borel-Lévy Formula) *Let λ denote the Lebesgue measure on I. For any $n \in \mathbb{N}_+$, the conditional probability $\lambda(T_N^n < x | \varepsilon_1, \ldots, \varepsilon_n)$ is given as follows:*

$$\lambda(T_N^n < x | \varepsilon_1, \ldots, \varepsilon_n) = \frac{(s_n^N + N)x}{s_n^N x + N}, \quad x \in I, \tag{3.2.3}$$

where s_n^N is as in (3.2.1) and $\varepsilon_1, \ldots, \varepsilon_n$ are as in (3.1.5).

Proof By definition, we have

$$\lambda\left(T_N^n < x | \varepsilon_1, \ldots, \varepsilon_n\right) = \frac{\lambda\left(\left(T_N^n < x\right) \cap I_N(\varepsilon_1, \ldots, \varepsilon_n)\right)}{\lambda\left(I_N(\varepsilon_1, \ldots, \varepsilon_n)\right)} \tag{3.2.4}$$

for any $n \in \mathbb{N}_+$ and $x \in I$. From (3.1.10) and (3.1.17) we have

$$\lambda\left(\left(T_N^n < x\right) \cap I_N(\varepsilon_1, \ldots, \varepsilon_n)\right) = \left| \frac{p_n}{q_n} - \frac{p_n + xp_{n-1}}{q_n + xq_{n-1}} \right| \tag{3.2.5}$$

$$= \frac{N^n x}{q_n(q_n + xq_{n-1})}.$$

From this and (3.1.16), we have

$$\lambda\left(T_N^n < x | \varepsilon_1, \ldots, \varepsilon_n\right) = \frac{\lambda\left(\left(T_N^n < x\right) \cap I_N(\varepsilon_1, \ldots, \varepsilon_n)\right)}{\lambda\left(I_N(\varepsilon_1, \ldots, \varepsilon_n)\right)}$$

3.2 Basic Metric Properties

$$= \frac{x(q_n + q_{n-1})}{q_n + xq_{n-1}} = \frac{(s_n^N + N)x}{s_n^N x + N} \quad (3.2.6)$$

for any $n \in \mathbb{N}_+$ and $x \in I$. □

The Brodén-Borel-Lévy formula allows us to determine the probability structure of incomplete quotients $(\varepsilon_n)_{n \in \mathbb{N}_+}$ under λ.

Proposition 3.2.2 *For any $i \geq N$ and $n \in \mathbb{N}_+$, we have*

$$\lambda(\varepsilon_1 = i) = \frac{N}{i(i+1)}, \quad \lambda(\varepsilon_{n+1} = i | \varepsilon_1, \ldots, \varepsilon_n) = V_{N,i}\left(s_n^N\right), \quad (3.2.7)$$

where $\{s_n^N\}$ is as in (3.2.1), and

$$V_{N,i}(x) := \frac{x+N}{(x+i)(x+i+1)}. \quad (3.2.8)$$

Proof From (3.1.15), the case $\lambda(\varepsilon_1 = i)$ holds. For $n \geq N$ and $x \in \Omega$, we have $T_N^n(x) = [\varepsilon_{n+1}, \varepsilon_{n+2}, \ldots]_N$ where $\{\varepsilon_n\}$ is as in (3.1.5). By using (3.2.3), we have

$$\lambda(\varepsilon_{n+1} = i | \varepsilon_1, \ldots, \varepsilon_n) = \lambda\left(T_N^n \in \left(\frac{N}{i+1}, \frac{N}{i}\right] | \varepsilon_1, \ldots, \varepsilon_n\right)$$

$$= \frac{(s_n^N + N)\frac{N}{i}}{s_n^N \frac{N}{i} + N} - \frac{(s_n^N + N)\frac{N}{i+1}}{s_n^N \frac{N}{i+1} + N} = V_{N,i}\left(s_n^N\right). \quad (3.2.9)$$

□

In (3.2.7), $\sum_{i \geq N} \lambda(\varepsilon_{n+1} = i | \varepsilon_1, \ldots, \varepsilon_n)$ must be 1 because λ is a probability measure on $(I, \mathcal{B}_I)$. This can be verified from (3.2.7) and (3.2.9) by using the partial fraction decomposition. By the same token, we see that

$$\sum_{i \geq N} V_{N,i}(x) = 1 \quad \text{for any } x \in I. \quad (3.2.10)$$

Remark 3.2.3 Proposition (3.2.2) is the starting point of an approach to the metrical theory of N-continued fraction expansions via dependence with complete connections (see [28], Section 5.2). We apply this method in Sect. 3.5.1 to obtain a solution of Gauss-Kuzmin-type problem for N-continued fraction expansions.

Corollary 3.2.4 *The sequence* $\left(s_n^N\right)_{n \in \mathbb{N}_+}$ *with* $s_0^N = 0$ *is a homogeneous* I*-valued Markov chain on* $(I, \mathcal{B}_I, \lambda)$ *with the following transition mechanism: From state* $s \in I$ *the only possible one-step transitions are those to states* $N/(s+i)$, $i \geq N$, *with corresponding probabilities* $V_{N,i}(s)$, $i \geq N$.

3.3 The Natural Extension of T_N and Extended Random Variables

Fix an integer $N \geq 1$. In this section, we introduce the natural extension $\overline{T}_N$ of T_N in (3.1.2) and its extended random variables [57].

3.3.1 Natural Extension of T_N

Let $(I, \mathcal{B}_I, T_N)$ be as in Definition 3.1.1(i). Define $(v_{N,i})_{i \geq N}$ by

$$v_{N,i} : I \to I; \quad v_{N,i}(x) := \frac{N}{x+i}, \quad x \in I. \quad (3.3.1)$$

For each $i \geq N$, $v_{N,i}$ is a right inverse of T_N, that is,

$$(T_N \circ v_{N,i})(x) = x, \quad \text{for any } x \in I. \quad (3.3.2)$$

Furthermore, if $\eta(x) = i$, then $(v_{N,i} \circ T_N)(x) = x$, where η is as in (3.1.4).

Definition 3.3.1 The natural extension $(I^2, \mathcal{B}_I^2, \overline{T}_N)$ of $(I, \mathcal{B}_I, T_N)$ is the transformation $\overline{T}_N$ of the square space $(I^2, \mathcal{B}_I^2) := (I, \mathcal{B}_I) \times (I, \mathcal{B}_I)$ defined as follows [49]:

$$\overline{T}_N : I^2 \to I^2; \quad \overline{T}_N(x, y) := \left(T_N(x), u_{N,\eta(x)}(y)\right), \quad (x, y) \in I^2. \quad (3.3.3)$$

From (3.3.2), we see that $\overline{T}_N$ is bijective on I^2 with the inverse

$$(\overline{T}_N)^{-1}(x, y) = (u_{N,\eta(y)}(x), T_N(y)), \quad (x, y) \in I^2. \quad (3.3.4)$$

Iterations of (3.3.3) and (3.3.4) are given as follows for each $n \geq 2$:

$$\left(\overline{T}_N\right)^n (x, y) = \left(T_N^n(x), [x_n, x_{n-1}, \ldots, x_2(x), x_1 + y]_N\right), \quad (3.3.5)$$

$$\left(\overline{T}_N\right)^{-n} (x, y) = \left([y_n, y_{n-1}, \ldots, y_2, y_1 + x]_N, T_N^n(y)\right), \quad (3.3.6)$$

where $x_i := \eta\left(T_N^{i-1}(x)\right)$ and $y_i := \eta\left(T_N^{i-1}(y)\right)$ for $i = 1, \ldots, n$.

3.3 The Natural Extension of T_N and Extended Random Variables

For G_N in (3.1.3), Dajani et al. [16] define its *extended measure* $\overline{G}_N$ on $(I^2, \mathcal{B}_I^2)$ as

$$\overline{G}_N(B) := \frac{1}{\log\left(\frac{N+1}{N}\right)} \iint_B \frac{N\,dx\,dy}{(xy+N)^2}, \quad B \in \mathcal{B}_I^2. \tag{3.3.7}$$

Then $\overline{G}_N(A \times I) = \overline{G}_N(I \times A) = G_N(A)$ for any $A \in \mathcal{B}_I$. The measure $\overline{G}_N$ is preserved by $\overline{T}_N$ [16].

3.3.2 Extended Random Variables

Define the projection $E : I^2 \to I$ by $E(x,y) := x$. With respect to $\overline{T}_N$ in (3.3.3), define *extended incomplete quotients* $\overline{\varepsilon}_l(x,y), l \in \mathbb{Z} := \{\ldots, -2, -1, 0, 1, 2, \ldots\}$ at $(x,y) \in I^2$ by

$$\overline{\varepsilon}_l(x,y) := (\eta \circ E)\left((\overline{T}_N)^{l-1}(x,y)\right), \quad l \in \mathbb{Z}. \tag{3.3.8}$$

Remark 3.3.2

(i) Remark that $\overline{\varepsilon}_l(x,y)$ in (3.3.8) is also well-defined for $l \leq 0$ because $\overline{T}_N$ is invertible. By (3.3.5) and (3.3.6), we have

$$\overline{\varepsilon}_n(x,y) = x_n, \quad \overline{\varepsilon}_0(x,y) = y_1, \quad \overline{\varepsilon}_{-n}(x,y) = y_{n+1}, \quad n \in \mathbb{N}_+, \ (x,y) \in I^2, \tag{3.3.9}$$

where we use notations in (3.3.5) and (3.3.6).

(ii) Since the measure $\overline{G}_N$ is preserved by $\overline{T}_N$, the doubly infinite sequence $(\overline{\varepsilon}_l(x,y))_{l \in \mathbb{Z}}$ is strictly stationary (i.e., its distribution is invariant under a shift of the indices) under $\overline{G}_N$.

Theorem 3.3.3 *Fix* $(x,y) \in I^2$ *and let* $\overline{\varepsilon}_l := \overline{\varepsilon}_l(x,y)$ *for* $l \in \mathbb{Z}$. *Define* $\varepsilon := [\overline{\varepsilon}_0, \overline{\varepsilon}_{-1}, \ldots]_N$. *Then the following holds for any* $x \in I$:

$$\overline{G}_N([0,x] \times I \mid \overline{\varepsilon}_0, \overline{\varepsilon}_{-1}, \ldots) = \frac{(N+\varepsilon)x}{\varepsilon x + N} \quad \overline{G}_N\text{-a.s.} \tag{3.3.10}$$

Proof Let $I_{N,n}$ denote the fundamental interval $I_N(\varepsilon_1, \varepsilon_2, \ldots, \varepsilon_n)$. Since $I_N(\overline{\varepsilon}_0, \overline{\varepsilon}_{-1}, \ldots, \overline{\varepsilon}_{-n}) = I \times I_{N,n+1}$ for $n \in \mathbb{N}$, we have

$$\overline{G}_N([0,x] \times I \mid \overline{\varepsilon}_0, \overline{\varepsilon}_{-1}, \ldots) = \lim_{n \to \infty} \overline{G}_N([0,x] \times I \mid \overline{\varepsilon}_0, \ldots, \overline{\varepsilon}_{-n}) \quad \overline{G}_N\text{-a.s.} \tag{3.3.11}$$

and

$$\overline{G}_N([0,x] \times I \mid \overline{\varepsilon}_0, \ldots, \overline{\varepsilon}_{-n}) = \frac{\overline{G}_N([0,x] \times I_{N,n+1})}{\overline{G}_N(I \times I_{N,n+1})}$$

$$= \frac{\left(\log\left(\frac{N+1}{N}\right)\right)^{-1}}{G_N(I_{N,n+1})} \int_{I_{N,n+1}} \mathrm{d}y \int_0^x \frac{N \mathrm{d}u}{(yu+N)^2}$$

$$= \frac{1}{G_N(I_{N,n+1})} \int_{I_{N,n+1}} \frac{x(y+N)}{xy+N} G_N(\mathrm{d}y)$$

$$= \frac{x(y_{n+1}+N)}{xy_{n+1}+N} \qquad (3.3.12)$$

for some $y_{n+1} \in I_{N,n+1}$. Since

$$\lim_{n\to\infty} y_{n+1} = [\overline{\varepsilon}_0, \overline{\varepsilon}_{-1}, \ldots]_N = \varepsilon, \qquad (3.3.13)$$

the proof is completed. $\square$

The stochastic property of $(\overline{\varepsilon}_l)_{l \in \mathbb{Z}}$ under $\overline{G}_N$ is given as follows.

Corollary 3.3.4 *For any $i \in \mathbb{N}_N$, we have*

$$\overline{G}_N(\overline{\varepsilon}_1 = i \mid \overline{\varepsilon}_0, \overline{\varepsilon}_{-1}, \ldots) = V_{N,i}(\varepsilon) \quad \overline{G}_N\text{-a.s.}, \qquad (3.3.14)$$

where $\varepsilon = [\overline{\varepsilon}_0, \overline{\varepsilon}_{-1}, \ldots]_N$ and $V_{N,i}$ is as in (3.2.8).

Proof Let $I_{N,n}$ be as in the proof of Theorem 3.3.3. Since $(\overline{\varepsilon}_1 = i) = I_N(i) \times I, i \in \mathbb{N}_N$, it follows that

$$\overline{G}_N(\overline{\varepsilon}_1 = i \mid \overline{\varepsilon}_0, \overline{\varepsilon}_{-1}, \ldots) = \lim_{n\to\infty} \overline{G}_N(\overline{\varepsilon}_1 = i \mid \overline{\varepsilon}_0, \overline{\varepsilon}_{-1}, \ldots, \varepsilon_{-n})$$

$$= \lim_{n\to\infty} \overline{G}_N(\overline{\varepsilon}_1 = i \mid I \times I_{N,n+1}) = \lim_{n\to\infty} \frac{\overline{G}_N\left(I_N(i) \times I_{N,n+1}\right)}{\overline{G}_N\left(I \times I_{N,n+1}\right)}$$

$$= \lim_{n\to\infty} \frac{1}{G_N(I_{N,n+1})} \int_{I_{N,n+1}} V_{N,i}(y) \, \mathrm{d}G_N(y)$$

$$= \lim_{n\to\infty} V_{N,i}(y_{n+1}) = V_{N,i}(\varepsilon) \quad \overline{G}_N\text{-a.s.},$$

where $y_{n+1} \in I_{N,n+1}$ with $\lim_{n\to\infty} y_{n+1} = [\overline{\varepsilon}_0, \overline{\varepsilon}_{-1}, \ldots]_N = \varepsilon$. $\square$

3.3 The Natural Extension of T_N and Extended Random Variables

Remark 3.3.5 The strict stationarity of $(\overline{\varepsilon}_l)_{l \in \mathbb{Z}}$ under $\overline{G}_N$ implies that

$$\overline{G}_N(\overline{\varepsilon}_{l+1} = i \mid \overline{\varepsilon}_l, \overline{\varepsilon}_{l-1}, \ldots) = V_{N,i}(\varepsilon) \quad \overline{G}_N\text{-a.s.} \tag{3.3.15}$$

for any $i \in \mathbb{N}_N$ and $l \in \mathbb{Z}$, where $\varepsilon := [\overline{\varepsilon}_l, \overline{\varepsilon}_{l-1}, \ldots]_N$. The last equation emphasizes that $(\overline{\varepsilon}_l)_{l \in \mathbb{Z}}$ is a chain of infinite order in the theory of dependence with complete connections [28].

Put $\overline{s}_\ell^N := [\overline{\varepsilon}_\ell^N, \overline{\varepsilon}_{\ell-1}^N, \ldots]_N$, $\ell \in \mathbb{Z}$. Note that $\overline{s}_\ell^N = N/(\overline{s}_{\ell-1}^N + \overline{\varepsilon}_\ell)$, $\ell \in \mathbb{Z}$. It follows from (3.3.15) that $\left(\overline{s}_\ell^N\right)_{\ell \in \mathbb{Z}}$ is a strictly stationary I-valued Markov process on $(I^2, \mathcal{B}_I^2, \overline{G}_N)$ with the following transition mechanism: From state $\overline{s} \in I$ the only possible transitions are those to states $N/(\overline{s} + i)$, $i \in \mathbb{N}_N$, the transition probability being $V_{N,i}(\overline{s})$. Clearly, whatever $\ell \in \mathbb{Z}$ we have

$$\overline{G}_N\left(\overline{s}_\ell^N < x\right) = \overline{G}_N\left(\overline{s}_0^N < x\right) = \overline{G}_N(I \times [0, x]) = G_N([0, x]), \quad x \in I.$$

Motivated by (3.3.10) we shall consider the family of (conditional) probability measures $(G_{N,t})_t$ on $\mathcal{B}_I$ defined by their distribution functions

$$G_{N,t}([0, x]) := \frac{(N + t)x}{tx + N}, \quad x, t \in I. \tag{3.3.16}$$

For any $t \in I$ put $s_{0,t}^N := t$ and $s_{n,t}^N := N/\left(s_{n-1,t}^N + \varepsilon_n\right)$, $n \in \mathbb{N}_+$. The motivation for considering the $s_{n,t}^N$, $n \in \mathbb{N}$, $t \in I$, will appear later on, see (3.3.20). Note that by the very definition of $s_{n,t}^N$, we have $s_{n,t}^N = [\varepsilon_n, \ldots, \varepsilon_2, \varepsilon_1 + t]_N$, $n \geq 2$, while $s_{1,t}^N = N/(\varepsilon_1 + t)$, $t \in I$.

Then $\left(s_{n,t}^N\right)_{n \in \mathbb{N}_+}$ is an I-valued Markov chain on $(I, \mathcal{B}_I, G_{N,t})$ which starts from $s_{t,0}^N = t$, $t \in I$, and has the following transition mechanism: From state $s \in I$ the only possible transitions are those to states $N/(s + i)$ with the corresponding transition probability $V_{N,i}(s)$, $i \in \mathbb{N}_N$. Let $B(I)$ denote the Banach space of all bounded $\mathcal{B}_I$-measurable complex-valued functions defined on I which is a Banach space under the supremum norm. The transition operator V_N of $\left(s_{n,t}^N\right)_{n \in \mathbb{N}_+}$ sends $f \in B(I)$ into the function defined by

$$E_t\left(f(s_{n+1,t}^N)\bigg| s_{n,t}^N = s\right) = \sum_{i \geq N} V_{N,i}(s) f\left(\frac{N}{s+i}\right) =: V_N f(s) \tag{3.3.17}$$

for any $s \in I$, where E_t stands for the mean value operator with respect to the probability measure $G_{N,t}$, whatever $t \in I$, and V_N coincides with the Perron-Frobenius operator of $(I, \mathcal{B}_I, G_N, T_N)$ (see Proposition 3.4.1(i)).

In connection with the operator V_N, if we define

$$V_N^\infty f := \int_I f(x) \mathrm{d} G_N(x), \quad f \in B(I), \tag{3.3.18}$$

then we have

$$V_N^\infty V_N^n f = V_N^\infty f, \text{ for any } f \in B(I) \text{ and } n \in \mathbb{N}_+. \tag{3.3.19}$$

Note that for any $t \in I$ and $n \in \mathbb{N}_+$ we have

$$G_{N,t}(A|\varepsilon_1, \ldots, \varepsilon_n) = G_{N, s_{n,t}^N}\left(T_N^n(A)\right),$$

whatever the set A belonging to the σ-algebra generated by the random variables $\varepsilon_{n+1}, \varepsilon_{n+2} \ldots$, that is, $\sigma(\varepsilon_{n+1}, \varepsilon_{n+2}, \ldots) = T_N^{-n}(\mathcal{B}_I)$. This follows from (3.3.10) for all irrational $t \in I$ and by continuity (use (3.3.16)) for all rational $t \in I$. In particular, it follows that the Brodén-Borel-Lévy formula holds under $G_{N,t}$ for any $t \in I$, that is,

$$G_{N,t}(T_N^n < x | \varepsilon_1, \ldots, \varepsilon_n) = \frac{(s_{n,t}^N + N)x}{s_{n,t}^N x + N}, \quad x \in I, n \in \mathbb{N}_+. \tag{3.3.20}$$

Note also that $\left(s_{n,t}^N\right)_{n \in \mathbb{N}}$ under $G_{N,t}$ is a version of $(\overline{s}_n^N)_{n \in \mathbb{N}}$ under $\overline{G}_N(\cdot | \overline{s}_0^N = t)$ for any $t \in I$.

3.4 Perron-Frobenius Operators

Let $(I, \mathcal{B}_I, G_N, T_N)$ be as in Definition 3.1.1(ii). In this section, we derive its Perron-Frobenius operator.

Let μ be a probability measure on $(I, \mathcal{B}_I)$ such that $\mu\left(T_N^{-1}(A)\right) = 0$ whenever $\mu(A) = 0$ for $A \in \mathcal{B}_I$. For example, this condition is satisfied if T_N is μ-preserving, that is, $\mu T_N^{-1} = \mu$. Let $L^1(I, \mu) := \{f : I \to \mathbb{C} : \int_I |f| \mathrm{d}\mu < \infty\}$. The *Perron-Frobenius operator* of $(I, \mathcal{B}_I, \mu, T_N)$ is defined as the bounded linear operator P_μ on the Banach space $L^1(I, \mu)$ such that the following holds:

$$\int_A P_\mu f \, \mathrm{d}\mu = \int_{T_N^{-1}(A)} f \, \mathrm{d}\mu \quad \text{for all } A \in \mathcal{B}_I, f \in L^1(I, \mu). \tag{3.4.1}$$

3.4 Perron-Frobenius Operators

Proposition 3.4.1 *Let $(I, \mathcal{B}_I, G_N, T_N)$ be as in Definition 3.1.1(ii), and let $V_N := P_{G_N}$ denote its Perron-Frobenius operator under G_N. Then the following holds:*

(i)

$$V_N f(x) = \sum_{i \geq N} V_{N,i}(x) f\left(\frac{N}{x+i}\right), \quad f \in L^1(I, G_N), \tag{3.4.2}$$

where $V_{N,i}$ is as in (3.2.8).

(ii) Let μ be a probability measure on $(I, \mathcal{B}_I)$ such that μ is absolutely continuous with respect to the Lebesgue measure λ and let $h := d\mu/d\lambda$ a.e. in I. Then the following holds:

(a) Let S_N denote the Perron-Frobenius operator of T_N under μ. Then the following holds a.e. in I:

$$S_N f(x) = \frac{N}{h(x)} \sum_{i \geq N} \frac{h\left(\frac{N}{x+i}\right)}{(x+i)^2} f\left(\frac{N}{x+i}\right) \tag{3.4.3}$$

$$= \frac{V_N \hat{f}_N(x)}{(x+N)h(x)} \tag{3.4.4}$$

for $f \in L^1(I, \mu)$, where $\hat{f}_N(x) := (x+N)f(x)h(x)$, $x \in I$. In addition, the n-th power S_N^n of S_N is written as follows:

$$S_N^n f(x) = \frac{V_N^n \hat{f}_N(x)}{(x+N)h(x)} \tag{3.4.5}$$

for any $f \in L^1(I, \mu)$ and any $n \geq 1$.

(b) Let L_N denote the Perron-Frobenius operator of T_N under λ. Then the following holds a.e. in I:

$$L_N f(x) = \sum_{i \geq N} \frac{N}{(x+i)^2} f\left(\frac{N}{x+i}\right), \quad f \in L^1(I, \lambda). \tag{3.4.6}$$

In addition, the n-th power L_N^n of L_N is written as follows:

$$L_N^n f(x) = \frac{V_N^n \bar{f}_N(x)}{x+N}, \quad f \in L^1(I, \lambda), \tag{3.4.7}$$

for any $f \in L^1(I, \lambda)$ and any $n \geq 1$, where $\bar{f}_N(x) := (x+N)f(x)$, $x \in I$.

(c) For any $n \in \mathbb{N}_+$ and $A \in \mathcal{B}_I$, we have

$$\mu\left(T_N^{-n}(A)\right) = \int_A V_N^n f_N(x) \mathrm{d}G_N(x), \qquad (3.4.8)$$

where $f_N(x) := \left(\log\left(\frac{N+1}{N}\right)\right)(x+N)h(x)$ for $x \in I$.

Proof (i) Let $T_{N,i}$ denote the restriction of T_N to the subinterval $I_{N,i} := \left(\frac{N}{i+1}, \frac{N}{i}\right], i \geq N$, that is,

$$T_{N,i}(x) = \frac{N}{x} - 1, \quad x \in I_i. \qquad (3.4.9)$$

Let $C(A) := T_N(A)$ and $C_i(A) := T_{N,i}^{-1}(A)$ for $A \in \mathcal{B}_I$. Since $C(A) = \bigcup_i C_i(A)$ and $C_i \cap C_j$ is a null set when $i \neq j$, we have

$$\int_{C(A)} f \, \mathrm{d}G_N = \sum_{i \geq N} \int_{C_i(A)} f \, \mathrm{d}G_N, \quad f \in L^1(I, G_N), \ A \in \mathcal{B}_I. \qquad (3.4.10)$$

For any $i \geq N$, by the change of variables $x = T_{N,i}^{-1}(y) = \frac{N}{y+i}$, we successively obtain

$$\int_{C_i(A)} f(x) \, G_N(\mathrm{d}x) = \left(\log\left(\frac{N+1}{N}\right)\right)^{-1} \int_{C_i(A)} \frac{f(x)}{N+x} \, \mathrm{d}x$$

$$= \left(\log\left(\frac{N+1}{N}\right)\right)^{-1} \int_A \frac{f\left(\frac{N}{y+i}\right)}{N + \frac{N}{y+i}} \frac{N}{(y+i)^2} \mathrm{d}y$$

$$= \int_A V_{N,i}(y) f\left(\frac{N}{y+i}\right) G_N(\mathrm{d}y). \qquad (3.4.11)$$

Now, (3.4.2) follows from (3.4.10) and (3.4.11).

(ii)(a) From (3.4.9), for any $f \in L^1(I, G_N)$ and $A \in \mathcal{B}_I$, we have

$$\int_{C(A)} f(x) \mu(\mathrm{d}x) = \sum_{i \geq N} \int_{C_i(A)} f(x) \mu(\mathrm{d}x). \qquad (3.4.12)$$

Then

$$\int_{C_i(A)} f(x) \mu(\mathrm{d}x) = \int_{C_i(A)} f(x) h(x) \, \mathrm{d}x$$

3.4 Perron-Frobenius Operators

$$= \int_A f\left(\frac{N}{y+i}\right) h\left(\frac{N}{y+i}\right) \frac{N}{(y+i)^2} \, dy. \tag{3.4.13}$$

From (3.4.12) and (3.4.13),

$$\int_{C(A)} f(x)\,\mu(dx) = \int_A \sum_{i \geq N} f\left(\frac{N}{x+i}\right) h\left(\frac{N}{x+i}\right) \frac{N}{(x+i)^2} \, dx. \tag{3.4.14}$$

Since $d\mu = h\,dx$, (3.4.3) follows from (3.4.14). Now, since $\hat{f}_N(x) = (x+N)f(x)h(x)$, from (3.4.3), we have

$$V_N \hat{f}_N(x) = N(x+N) \sum_{i \geq N} \frac{h\left(\frac{N}{x+i}\right)}{(x+i)^2} f\left(\frac{N}{x+i}\right). \tag{3.4.15}$$

From (3.4.3) and (3.4.15), (3.4.4) follows immediately.
(ii)(b) The formula (3.4.6) is a consequence of (3.4.4) and follows immediately.
(ii)(c) We will use mathematical induction. For $n = 0$, Eq. (3.4.8) holds by definitions of f_N and h. Assume that (3.4.8) holds for some $n \in \mathbb{N}$. Then

$$\mu\left(T_N^{-n-1}(A)\right) = \mu\left(T_N^{-n}\left(T_N^{-1}(A)\right)\right) = \int_{C(A)} V_N^n f_N(x)\, G_N(dx). \tag{3.4.16}$$

Since V_N is the Perron-Frobenius operator under G_N, by (3.4.1), we have

$$\int_{C(A)} V_N^n f_N(x)\, G_N(dx) = \int_A V_N^{n+1} f_N(x)\, G_N(dx). \tag{3.4.17}$$

Therefore,

$$\mu\left(T_N^{-n-1}(A)\right) = \int_A V_N^{n+1} f_N(x) G_N(dx), \tag{3.4.18}$$

which ends the proof. □

In the following proposition we show that the operator V_N in (3.4.2) preserves monotonicity and enjoys a contraction property for Lipschitz continuous functions.

Proposition 3.4.2 *Let V_N be as in (3.4.2):*

(i) *Let $f \in B(I)$. Then the following holds:*
  (a) *If f is nondecreasing (nonincreasing), then $V_N f$ is nonincreasing (nondecreasing).*

(b) If f is monotone, then

$$\text{var}(V_N f) \leq \frac{1}{N+1} \cdot \text{var} f + \widetilde{K}_N \cdot |f|, \qquad (3.4.19)$$

where

$$\widetilde{K}_N := \frac{2\sqrt{N^2 + N}}{(N + \sqrt{N^2 + N})(N + \sqrt{N^2 + N} + 1)}. \qquad (3.4.20)$$

(ii) For any $f \in L(I)$, we have

$$s(V_N f) \leq \widetilde{q}_N \cdot s(f), \qquad (3.4.21)$$

where

$$\widetilde{q}_N := N \cdot \sum_{i \geq N} \left(\frac{N}{i^3(i+1)} + \frac{i+1-N}{i(i+1)^3} \right). \qquad (3.4.22)$$

Proof See [39] and [66]. □

Corollary 3.4.3 *For any $f \in BV(I)$ and for all $n \in \mathbb{N}$ we have*

$$\text{var} V_N^n f \leq \left(\frac{1}{N+1} + \widetilde{K}_N \right)^n \cdot \text{var} f, \qquad (3.4.23)$$

$$\left| V_N^n f - V_N^\infty f \right| \leq \left(\frac{1}{N+1} + \widetilde{K}_N \right)^n \cdot \text{var} f. \qquad (3.4.24)$$

Proof See [66]. □

3.5 Gauss-Kuzmin Theorem for N-Continued Fractions

3.5.1 The One-Dimensional Case

The problem of finding the asymptotic of $T_N^{-n}(A)$ as $n \to \infty$ represents the Gauss-Kuzmin-type problem for N-continued fraction expansions.

Theorem 3.5.1 (A Gauss-Kuzmin Theorem for T_N) *Fix an integer $N \geq 1$ and let $(I, \mathcal{B}_I, T_N)$ be as above:*

3.5 Gauss-Kuzmin Theorem for N-Continued Fractions

(i) For a probability measure μ on $(I, \mathcal{B}_I)$, let the assumption (A) as follows:

(A) μ *is nonatomic and has a Riemann-integrable density.*

Then for any probability measure μ which satisfies (A), the following holds:

$$\lim_{n \to \infty} \mu(T_N^n < x) = \frac{1}{\log\{(N+1)/N\}} \log \frac{x+N}{N}, \quad x \in I. \tag{3.5.1}$$

(ii) In addition to assumptions of μ in (i), if the density of $I \ni x \mapsto \mu([0, x])$ is Lipschitz continuous, then there exist two positive constants $\ell_N < 1$ and $\widetilde{k}_N$ and such that for any $x \in I$ and $n \geq 1$, there exists θ_N with $|\theta_N| \leq \widetilde{k}_N$, the following holds:

$$\mu\left(T_N^n < x\right) = \frac{1 + \theta_N \ell_N^n}{\log\{(N+1)/N\}} \log \frac{x+N}{N}, \tag{3.5.2}$$

where $\theta_N := \theta_N(N, \mu, n, x)$. As a consequence, the n-th error term $e_{N,n}(\mu; x)$ of the Gauss-Kuzmin problem is obtained as follows:

$$e_{N,n}(\mu; x) = \frac{\theta_N \ell_N^n}{\log\{(N+1)/N\}} \log \frac{x+N}{N}. \tag{3.5.3}$$

In order to solve the problem, we apply the theory of random systems with complete connections by Iosifescu [28].

Fix an integer $N \geq 1$. We introduce a random system with complete connections (=RSCC) $\{(I, \mathcal{B}_I), (\mathbb{N}_N, \mathcal{P}(\mathbb{N}_N)), v_N, \widetilde{V}_N\}$ as follows:

$$\begin{cases} v_N : I \times \mathbb{N}_N \to I; & v_N(x, i) := v_{N,i}(x) = \dfrac{N}{x+i}, \\ \widetilde{V}_N : I \times \mathbb{N}_N \to I; & \widetilde{V}_N(x, i) := V_{N,i}(x) = \dfrac{x+N}{(x+i)(x+i+1)}, \\ \mathbb{N}_N := \{N, N+1, \ldots\}; \ \mathcal{P}(\mathbb{N}_N) = \text{ the power set of } \mathbb{N}_N. \end{cases} \tag{3.5.4}$$

By the definition of $\widetilde{V}_N$ and using the partial fraction decomposition, $\sum_{i \geq N} \widetilde{V}_N(x, i) = 1$.

By Sect. 1.4,

$$\widetilde{Q}_N(x, B) = \sum_{i \in B_{N,x}} V_{N,i}(x) \tag{3.5.5}$$

for $(x, B) \in I \times \mathcal{B}_I$, where $B_{N,x} := \{i \in \mathbb{N}_N : v_{N,i}(x) \in B\}$. Let V_N be the transition operator associated with the Markov chain $\left(s_{n,t}^N\right)_{n \in \mathbb{N}_+}$ with state space $(I, \mathcal{B}_I)$ and transition probability function $\widetilde{Q}_N$.

For the dynamical system $(I, \mathcal{B}_I, T_N)$ in 3.1.1 and a given probability measure μ on $(I, \mathcal{B}_I)$, the ergodic behavior of the RSCC in (3.5.4) allows us to find the limiting Gauss-Kuzmin distribution $\widetilde{F}_N$ with respect to (T_N, μ):

$$\widetilde{F}_N(x) := \lim_{n \to \infty} \mu(T_N^n < x), \quad x \in I \tag{3.5.6}$$

and the invariant measure induced by $\widetilde{F}_N$.

Proposition 3.5.2 *The RSCC defined by (3.5.4) is with contraction.*

Proof We have

$$\frac{d}{dx} v_{N,i}(x) = -\frac{N}{(x+i)^2},$$

$$\frac{d}{dx} V_{N,i}(x) = \frac{i^2 + i - 2Ni - (x^2 + 2Nx + N)}{(x+i)^2(x+i+1)^2}$$

for any $x \in I$ and $i \geq N$. Thus,

$$\sup_{x \in I} \left| \frac{d}{dx} v_{N,i}(x) \right| \leq \frac{N}{i^2}, \quad i \geq N$$

$$\sup_{x \in I} \left| \frac{d}{dx} V_{N,i}(x,i) \right| < \infty.$$

Hence the requirements of Definition 1.4.8 are fulfilled. □

Proposition 3.5.3 *The RSCC $\{(I, \mathcal{B}_I), (\mathbb{N}_N, \mathcal{P}(\mathbb{N}_N)), v_N, \widetilde{V}_N\}$ has a regular associated Markov chain.*

Proof By Theorem 1.4.9(i), it is equivalent to that

$$(\Gamma): \left\{ \begin{array}{l} \text{there exists a point } x^* \in I \text{ such that} \\ \lim_{n \to \infty} \text{dist}(\sigma_n(x), x^*) = 0 \quad \text{for all } x \in I, \end{array} \right\}$$

where we remark that $W = I = [0, 1]$ in this case. Hence we show the condition (Γ) as follows.

Fix $x \in I$. Let us define the sequence $(x_n)_{n \geq 0}$ in I, recursively by $x_0 := x$, $x_{n+1} := \dfrac{N}{x_n + N}$ for $n \geq 1$. Clearly $x_{n+1} \in \sigma_1(x_n)$ and therefore Theorem 1.4.9(ii) and an induction argument lead us to the conclusion that $x_n \in \sigma_n(x)$ for $n \in \mathbb{N}_+$. But, $\lim_{n \to \infty} x_n =$

3.5 Gauss-Kuzmin Theorem for N-Continued Fractions

$x^* = \dfrac{-N + \sqrt{N^2 + 4N}}{2}$ for any $x \in I$. Hence $\text{dist}(\sigma_n(x), x^*) \leq |x_n - x^*| \to 0$ as $n \to \infty$. Hence we find $x^* := \dfrac{-N + \sqrt{N^2 + 4N}}{2}$ in the condition (Γ). $\square$

Proposition 3.5.4 *The system $\{(I, \mathcal{B}_I), (\mathbb{N}_N, \mathcal{P}(\mathbb{N}_N)), v_N, \widetilde{V}_N\}$ is an ergodic RSCC.*

Proof Since the above RSCC has a regular associated Markov chain, then the statement holds. $\square$

Let $L(I)$ denote the Banach space of all complex-valued Lipschitz continuous functions on I. Since the compact Markov chain corresponding to this RSCC is ergodic with respect to $L(I)$, Theorem 1.4.6(ii) allows us to obtain the unique stationary probability measure $\widetilde{Q}_N^\infty$ on $(I, \mathcal{B}_I)$.

Proposition 3.5.5 *The probability $\widetilde{Q}_N^\infty$ is the invariant probability measure of the transformation T_N.*

Proof For G_N in Definition 3.1.1 and $\widetilde{Q}_N$ in (3.5.5), and on account of the uniqueness of $\widetilde{Q}_N^\infty$, we have to show that

$$\int_0^1 \widetilde{Q}_N(x, B) \, dG_N(x) = G_N(B), \quad B \in \mathcal{B}_I. \tag{3.5.7}$$

Since the intervals $[0, u) \subset [0, 1)$ generate $\mathcal{B}_I$, it is sufficient to show Eq. (3.5.7) just for $B = [0, u), 0 < u \leq 1$. Let $E(x, N) = \lfloor \frac{N}{u} - x \rfloor + 1$. By (3.5.5), we have

$$\widetilde{Q}_N(x, [0, u)) = \sum_{\{i \in \mathbb{N}_N : 0 \leq v_{N,i}(x) < u\}} V_{N,i}(x) = \sum_{i \geq E(x, N)} V_{N,i}(x)$$

$$= \frac{N - E(x, N)}{x + E(x, N)}. \tag{3.5.8}$$

Thus,

$$\int_0^1 \widetilde{Q}_N(x, [0, u)) dG_N(x) = \frac{1}{\log\{(N+1)/N\}} \log \frac{x + N}{N} = G_N([0, u)). \tag{3.5.9}$$

Hence the statement holds. $\square$

The regularity of V_N in (3.4.2) with respect to $L(I)$ implies that V_N is aperiodic. By Lemma 1.4.7 there exist two positive constants $\ell_N < 1$ and $\widetilde{k}_N$ such that

$$\|V_N^n f - V_N^\infty f\|_L \leq \widetilde{k}_N \ell_N^n \|f\|_L, \quad n \in \mathbb{N}_+, \; f \in L(I), \tag{3.5.10}$$

where

$$V_N^n : L(I) \to L(I); \quad V_N^n f(x) = \int_I f(y) \, \widetilde{Q}_N^n(x, \mathrm{d}y), \tag{3.5.11}$$

$$V_N^\infty : L(I) \to \mathbb{C}; \quad V_N^\infty f = \int_I f(y) \, \widetilde{Q}_N^\infty(\mathrm{d}y). \tag{3.5.12}$$

Proof of Theorem 3.5.1 By (3.4.8), we have

$$\mu\left(T_N^{-n}(A)\right) = \int_A \frac{V_N^n f_{N,0}(x)}{x+N} \mathrm{d}x \quad \text{for any } n \in \mathbb{N}_+, \; A \in \mathcal{B}_I, \tag{3.5.13}$$

where $f_{N,0}(x) = (x+N)(\mathrm{d}\mu/\mathrm{d}\lambda)(x)$ for $x \in I$. If $\mathrm{d}\mu/\mathrm{d}\lambda \in L(I)$, then $f_{N,0} \in L(I)$, and by (3.5.12) we have

$$V_N^\infty f_{N,0} = \int_I f_{N,0}(x) \, \widetilde{Q}_N^\infty(\mathrm{d}x) = \int_I f_{N,0}(x) \, G_N(\mathrm{d}x)$$

$$= \frac{1}{\log\{(N+1)/N\}}. \tag{3.5.14}$$

Taking into account (3.5.10), there exist two constants $\ell_N < 1$ and $\widetilde{k}_N$ such that

$$\|V_N^n f_{N,0} - V_N^\infty f_{N,0}\|_L \leq \widetilde{k}_N \ell_N^n \|f_{N,0}\|_L, \quad n \in \mathbb{N}_+. \tag{3.5.15}$$

Furthermore, consider the Banach space $C(I)$ of all real-valued continuous functions on I with the norm $\|f\| := \sup_{x \in I} |f(x)|$. Since $L(I)$ is a dense subspace of $C(I)$ we have

$$\lim_{n \to \infty} \|(V_N^n - V_N^\infty) f\| = 0 \quad \text{for all } f \in C(I). \tag{3.5.16}$$

Therefore, (3.5.16) is valid for any measurable function f which is $\widetilde{Q}_N^\infty$-almost surely continuous, that is, for a Riemann-integrable function. Thus, we have

$$\lim_{n \to \infty} \mu\left(T_N^n < x\right) = \lim_{n \to \infty} \int_0^x \frac{V_N^n f_{N,0}(u)}{u+N} \mathrm{d}u = \int_0^x \frac{V_N^\infty f_{N,0}(u)}{u+N} \mathrm{d}u$$

$$= \frac{1}{\log\{(N+1)/N\}} \int_0^x \frac{\mathrm{d}u}{u+N}$$

3.5 Gauss-Kuzmin Theorem for N-Continued Fractions

$$= \frac{1}{\log\{(N+1)/N\}} \log \frac{x+N}{N}.$$

Hence (3.5.1) is proved. □

3.5.2 A Two-Dimensional Case

In this section we give a Gauss-Kuzmin theorem related to the natural extension $(I^2, \mathcal{B}_I^2, \overline{G}_N, \overline{T}_N)$. To solve the problem, we need a version of the Gauss-Kuzmin theorem for T_N different from that presented in the previous section.

Theorem 3.5.6 (Another Version of Gauss-Kuzmin Theorem for T_N) *Let (I, G_N, T_N) as in Definition 3.1.1. If λ is the Lebesgue measure on I, then there exists a constant $0 < \ell_N < 1$ such that for any $A \in \mathcal{B}_I$ we have*

$$\left|\lambda\left(T_N^{-n}(A)\right) - G_N(A)\right| < C_N \ell_N^n \lambda(A), \qquad (3.5.17)$$

where C_N is a universal constant.

Proof In Proposition 3.4.1(ii)(c) we showed that $\mu\left(T_N^{-n}(A)\right) = \int_A V_N^n f_N(x) \mathrm{d}G_N(x)$, where μ is a probability measure on $(I, \mathcal{B}_I)$ absolutely continuous with respect to the Lebesgue measure λ, and $f_N(x) := \left(\log\left(\frac{N+1}{N}\right)\right)(x+N)h(x)$ with $h := \mathrm{d}\mu/\mathrm{d}\lambda$ a.e. in I. In the special case $\mu = \lambda$ we obviously have

$$\lambda\left(T_N^{-n}(A)\right) = \frac{1}{\log\left(\frac{N+1}{N}\right)} \int_A \frac{V_N^n f_N(x)}{x+N} \mathrm{d}x \qquad (3.5.18)$$

with $f_N(x) := \left(\log\left(\frac{N+1}{N}\right)\right)(x+N), x \in I$. Thus, from (3.5.14) we have that $V_N^\infty f_N = 1$. Therefore,

$$G_N(A) = \frac{1}{\log\left(\frac{N+1}{N}\right)} \int_A \frac{V_N^\infty f_N(x)}{x+N} \mathrm{d}x. \qquad (3.5.19)$$

In (3.5.10) we showed that there are two positive constants $\ell_N < 1$ and $\widetilde{k}_N$ such that

$$\left\|V_N^n g - V_N^\infty g\right\|_L \leq \widetilde{k}_N \ell_N^n \|g\|_L, \quad g \in L(I), n \in \mathbb{N}_+, \qquad (3.5.20)$$

where $L(I)$ denotes the Banach space of all complex-valued Lipschitz continuous functions on I with the following norm:

$$\|g\|_L := \sup_{x \in I} |g(x)| + \sup_{x' \neq x''} \frac{|g(x') - g(x'')|}{|x' - x''|}.$$

Therefore

$$\left|\lambda\left(T_N^{-n}(A)\right) - G_N(A)\right| \leq \frac{1}{\log\left(\frac{N+1}{N}\right)} \int_A \frac{|V_N^n f_N(x) - V_N^\infty f_N(x)|}{x+N} dx$$

$$< \widetilde{k}_N \ell_N^n \|f_N\|_L \frac{1}{\log\left(\frac{N+1}{N}\right)} \int_A \frac{1}{x+N} dx$$

$$= \widetilde{k}_N \ell_N^n \|f_N\|_L G_N(A).$$

Since

$$G_N(A) \leq \frac{1}{N \log\left(\frac{N+1}{N}\right)} \lambda(A), \quad A \in \mathcal{B}_I,$$

the proof is completed. □

The essential argument in the proof of the Gauss-Kuzmin theorem for $\overline{T}_N$ is the Gauss-Kuzmin equation. First we give the Gauss-Kuzmin equation for the one-dimensional case. Define the functions $\left(\widetilde{F}_{N,n}\right)_{n \in \mathbb{N}}$ on I by

$$\widetilde{F}_{N,0}(x) := x, \quad \widetilde{F}_{N,n}(x) := \lambda(T_N^n < x), n \geq 1. \tag{3.5.21}$$

The Gauss-Kuzmin equation is

$$\widetilde{F}_{N,n+1}(x) = \sum_{i \geq N} \left\{ \widetilde{F}_{N,n}\left(\frac{N}{i}\right) - \widetilde{F}_{N,n}\left(\frac{N}{x+i}\right) \right\} \tag{3.5.22}$$

for $x \in I$ and $n \in \mathbb{N}$. The density of the measure G_N defined in (3.1.3) is an eigenfunction of (3.5.22), namely, if we put $\widetilde{F}_{N,n}(x) = \log\left(\frac{x+N}{N}\right)$, $x \in I$, we obtain $\widetilde{F}_{N,n+1}(x) = \log\left(\frac{x+N}{N}\right)$.

Next we will derive a Gauss-Kuzmin theorem related to the natural extension $(I^2, \mathcal{B}_I^2, \overline{G}_N, \overline{T}_N)$ defined in Sect. 3.3.

3.5 Gauss-Kuzmin Theorem for N-Continued Fractions

For any $n \in \mathbb{N}_+$ and $x, y \in I$, let us define $\Delta_{x,y} := [0, x] \times [0, y]$ and the functions $\overline{F}_{N,n}(x, y)$ by

$$\overline{F}_{N,n}(x, y) := \overline{\lambda}\left((\overline{T}_N)^n \in \Delta_{x,y}\right), \qquad (3.5.23)$$

where $\overline{\lambda}$ is the Lebesgue measure on I^2. Then the following holds.

Theorem 3.5.7 (A Gauss-Kuzmin Theorem for $\overline{T}_N$) *For every $n \geq 2$ and $(x, y) \in I^2$ one has*

$$\overline{F}_{N,n}(x, y) = \overline{G}_N(\Delta_{x,y}) = \frac{1}{\log\left(\frac{N+1}{N}\right)} \log\left(\frac{xy+N}{N}\right) + \mathcal{O}\left(\overline{\alpha}_N\right)^n \qquad (3.5.24)$$

with $0 < \overline{\alpha}_N < 1$.

For any $0 < y \leq 1$, put $\ell_1 := \left\lfloor \frac{N}{y} \right\rfloor$. For $(\xi, \zeta) \in I^2$, then $(\overline{T}_N)^{n+1}(\xi, \zeta) \in \Delta_{x,y}$ is equivalent to

$$(\overline{T}_N)^n \in \left(\bigcup_{i \geq \ell_1 + 1} \left[\frac{N}{x+i}, \frac{N}{i}\right] \times [0, 1]\right) \cup \left(\left[\frac{N}{x+\ell_1}, \frac{N}{\ell_1}\right] \times \left[\frac{N}{y} - \ell_1, 1\right]\right).$$

Now, from (3.5.23) we have the following recursive formula that represents the Gauss-Kuzmin equation associated with the functions $(\overline{F}_{N,n})_{n \in \mathbb{N}_+}$:

$$\overline{F}_{N,n+1}(x, y) = \sum_{i \geq \ell_1} \left\{ \overline{F}_{N,n}\left(\frac{N}{i}, 1\right) - \overline{F}_{N,n}\left(\frac{N}{x+i}, 1\right) \right\} \qquad (3.5.25)$$

$$- \left\{ \overline{F}_{N,n}\left(\frac{N}{\ell_1}, \frac{N}{y} - \ell_1\right) - \overline{F}_{N,n}\left(\frac{N}{x+\ell_1}, \frac{N}{y} - \ell_1\right) \right\}.$$

The density of the measure $\overline{G}_N$ defined in (3.1.3) is an eigenfunction of (3.5.25), namely, if we put $\overline{F}_{N,n}(x, y) = \log\left(\frac{xy+N}{N}\right)$, $x, y \in I$, we obtain $\overline{F}_{N,n+1}(x, y) = \log\left(\frac{xy+N}{N}\right)$.

Lemma 3.5.8 *Let $n \in \mathbb{N}$, $n \geq 2$ and let $y \in I \cap \mathbb{Q}$ with $y = [\ell_1, \ldots, \ell_d]_N$, $\ell_1, \ldots, \ell_d \in \mathbb{N}_N$, $\ell_d \geq N+1$, where $d \leq \lfloor n/(N+1) \rfloor$. Then for every $x, x^* \in [0, 1)$ with $x^* < x$,*

$$\left| \overline{F}_{N,n}(x, y) - \overline{F}_{N,n}(x^*, y) - \frac{\log\left(\frac{xy+N}{x^*y+N}\right)}{\log\left(\frac{N+1}{N}\right)} \right| < \frac{(N+1)^3}{N^2} \frac{C_N}{1 - \ell_N} \overline{\lambda}(\Delta_{x,y} \setminus \Delta_{x^*,y}) \ell_N^{n-d},$$

where $0 < \ell_N < 1$ and C_N are given as in Theorem 3.5.6.

Proof See [66]. □

Proof of Theorem 3.5.7 Let $(x, y) \in I^2$, $n \geq 2$. In view of Lemma 3.5.8 we assume that $y \notin \mathbb{Q}$. Put $d := \max \{\kappa \in \mathbb{N} : \kappa \text{ even and } \kappa \leq \lfloor n/(N+1) \rfloor + 2\}$.

Let q_n be as in (3.1.8). Since $\varepsilon_n \geq N$, $n \in \mathbb{N}_+$, we have $q_n \geq N(q_{n-1} + q_{n-2})$, $n \in \mathbb{N}_+$. Define now $\overline{q}_n$, $n \in \mathbb{N}_+$, by the recurrence relation

$$\overline{q}_n := N(\overline{q}_{n-1} + \overline{q}_{n-2}), n \in \mathbb{N}_+, \quad (3.5.26)$$

with $\overline{q}_{-1} = 0$ and $\overline{q}_0 = 1$. Then

$$q_n \geq \overline{q}_n, n \in \mathbb{N}_+. \quad (3.5.27)$$

It is easy to see that

$$\overline{q}_n = \frac{1}{\sqrt{N^2 + 4N}} \left\{ \left(\frac{N + \sqrt{N^2 + 4N}}{2} \right)^{n+1} - \left(\frac{N - \sqrt{N^2 + 4N}}{2} \right)^{n+1} \right\}. \quad (3.5.28)$$

From (3.1.12), (3.5.27), and (3.5.28) and the fact that d is even and $\frac{n}{N+1} < d \leq \frac{n}{N+1} + 2$, we get

$$\left| y - \frac{p_d}{q_d} \right| < \frac{N^d}{(\overline{q}_d)^2} = \frac{N^d (N^2 + 4N)}{\left\{ \left(\frac{N+\sqrt{N^2+4N}}{2} \right)^{d+1} - \left(\frac{N-\sqrt{N^2+4N}}{2} \right)^{d+1} \right\}^2}$$

$$< \frac{N^d (N^2 + 4N)}{\left\{ \left(\frac{N+\sqrt{N^2+4N}}{2} \right)^{d+1} \right\}^2} < \frac{(N)^{\frac{n}{N+1}+2} \cdot (N^2 + 4N) \cdot 2^{\frac{2n}{N+1}+2}}{\left(N + \sqrt{N^2 + 4N} \right)^{\frac{2n}{N+1}+2}}. \quad (3.5.29)$$

Now for each $B \in \mathcal{B}_I^2$ one has

$$\frac{N}{(N+1)^2 \log \left(\frac{N+1}{N} \right)} \overline{\lambda}(B) \leq \overline{G}_N(B) \leq \frac{1}{N \log \left(\frac{N+1}{N} \right)} \overline{\lambda}(B). \quad (3.5.30)$$

Since $\Delta_{x, p_d/q_d} \subset \Delta_{x,y}$ and $\overline{F}_{N,n}(x, y) = \overline{\lambda} \left(\left(\overline{T}_N \right)^{-n} (\Delta_{x,y}) \right)$, from (3.5.30), (3.5.29) and the fact that $\overline{T}_N$ is $\overline{G}_N$-invariant, we find that

3.5 Gauss-Kuzmin Theorem for N-Continued Fractions

$$\overline{F}_{N,n}(x,y) - \overline{F}_{N,n}(x, \tfrac{p_d}{q_d}) = \overline{\lambda}\left((\overline{T}_N)^{-n}(\Delta_{x,y}) \setminus (\overline{T}_N)^{-n}(\Delta_{x,p_d/q_d})\right)$$

$$\leq \frac{(N+1)^3}{N^2} \log\left(\frac{N+1}{N}\right) \overline{G}_N\left((\overline{T}_N)^{-n}(\Delta_{x,y}) \setminus (\overline{T}_N)^{-n}(\Delta_{x,p_d/q_d})\right)$$

$$\leq \left(\frac{N+1}{N}\right)^3 \overline{\lambda}\left([0,x] \times \left[\tfrac{p_d}{q_d}, y\right]\right) \leq \left(\frac{N+1}{N}\right)^3 x \left|y - \frac{p_d}{q_d}\right|$$

$$< \left(\frac{N+1}{N}\right)^3 \frac{(N)^{\frac{n}{N+1}+2} \cdot (N^2+4N) \cdot 2^{\frac{2n}{N+1}+2} \cdot x}{\left(N + \sqrt{N^2+4N}\right)^{\frac{2n}{N+1}+2}}.$$

(3.5.31)

Since for every fixed $x \in [0,1]$ the function $y \mapsto \log\left(\frac{xy+N}{N}\right)$ is differentiable on $[0,1]$, by the *Mean Value Theorem* we have

$$\left|\log\left(\frac{xy+N}{N}\right) - \log\left(\frac{x\tfrac{p_d}{q_d}+N}{N}\right)\right| = \left|y - \frac{p_d}{q_d}\right| \cdot \left|\frac{x}{x\xi + N}\right|$$

$$\leq x \cdot \left|y - \frac{p_d}{q_d}\right| < \frac{(N)^{\frac{n}{N+1}+2} \cdot (N^2+4N) \cdot 2^{\frac{2n}{N+1}+2} \cdot x}{\left(N + \sqrt{N^2+4N}\right)^{\frac{2n}{N+1}+2}},$$

(3.5.32)

where $p_d/q_d \leq \xi \leq y$.

From Lemma 3.5.8, (3.5.31), and (3.5.32), and the fact $\overline{F}_{N,n}\left(0, \tfrac{p_d}{q_d}\right) = 0$, we have

$$\left|\overline{F}_{N,n}(x,y) - \frac{1}{\log\left(\frac{N+1}{N}\right)} \log\left(\frac{xy+N}{N}\right)\right| \leq \left|\overline{F}_{N,n}(x,y) - \overline{F}_{N,n}\left(x, \tfrac{p_d}{q_d}\right)\right|$$

$$+ \left|\overline{F}_{N,n}\left(x, \tfrac{p_d}{q_d}\right) - \overline{F}_{N,n}\left(0, \tfrac{p_d}{q_d}\right) - \frac{1}{\log\left(\frac{N+1}{N}\right)} \log\left(\frac{x\tfrac{p_d}{q_d}+N}{N}\right)\right|$$

$$+ \frac{1}{\log\left(\frac{N+1}{N}\right)} \left|\log\left(\frac{xy+N}{N}\right) - \log\left(\frac{x\tfrac{p_d}{q_d}+N}{N}\right)\right|$$

$$\leq \left(\frac{N+1}{N}\right)^3 \frac{(N)^{\frac{n}{N+1}+2} \cdot (N^2+4N) \cdot 2^{\frac{2n}{N+1}+2} \cdot x}{\left(N + \sqrt{N^2+4N}\right)^{\frac{2n}{N+1}+2}} + \frac{(N+1)^3}{N^2} \frac{C_N}{1-\ell_N} x \frac{p_d}{q_d} \ell_N^{n-d}$$

$$+\frac{1}{\log\left(\frac{N+1}{N}\right)}\frac{(N)^{\frac{n}{N+1}+2}\cdot(N^2+4N)\cdot 2^{\frac{2n}{N+1}+2}\cdot x}{\left(N+\sqrt{N^2+4N}\right)^{\frac{2n}{N+1}+2}}$$

$$\leq\left(\left(\frac{N+1}{N}\right)^3+\frac{1}{\log\left(\frac{N+1}{N}\right)}\right)\frac{(N)^{\frac{n}{N+1}+2}\cdot(N^2+4N)\cdot 2^{\frac{2n}{N+1}+2}}{\left(N+\sqrt{N^2+4N}\right)^{\frac{2n}{N+1}+2}}$$

$$+\frac{(N+1)^3}{N^2}\frac{C_N}{1-\ell_N}\ell_N^{n-d}.$$

Since $d \leq n/(N+1)+2$, we have

$$\left|\overline{F}_{N,n}(x,y)-\frac{\log\left(\frac{xy+N}{N}\right)}{\log\left(\frac{N+1}{N}\right)}\right|\leq\left(\left(\frac{N+1}{N}\right)^3+\frac{1}{\log\left(\frac{N+1}{N}\right)}\right)$$

$$\times\frac{(N)^{\frac{n}{N+1}+2}\cdot(N^2+4N)\cdot 2^{\frac{2n}{N+1}+2}\cdot x}{\left(N+\sqrt{N^2+4N}\right)^{\frac{2n}{N+1}+2}}$$

$$+\frac{(N+1)^3}{N^2}\frac{C_N}{(1-\ell_N)\ell_N^2}\ell_N^{n\frac{N}{N+1}}\leq\mathcal{K}_N\left(\overline{\alpha}_N\right)^n,$$

where

$$\mathcal{K}_N:=\frac{4N^2(N^2+4N)}{(N+\sqrt{N^2+4N})^2}\left(\left(\frac{N+1}{N}\right)^3+\frac{1}{\log\left(\frac{N+1}{N}\right)}\right)+\frac{(N+1)^3}{N^2}\frac{C_N}{(1-\ell_N)\ell_N^2}$$

and

$$\overline{\alpha}_N:=\max\left(\ell_N^{\frac{N}{N+1}},\frac{(4N)^{\frac{1}{N+1}}}{(N+\sqrt{N^2+4N})^{\frac{2}{N+1}}}\right)<1.$$

The proof is completed. □

3.5.3 Lower and Upper Bounds for the Convergence Rate of the Distribution Function

In this section we shall estimate the error term

3.5 Gauss-Kuzmin Theorem for N-Continued Fractions

$$e_{N,n,t}(x, y) = G_{N,t}\left(T_N^n \in [0, x], s_{n,t}^N \in [0, y]\right) - \frac{1}{\log\left(\frac{N+1}{N}\right)} \log\left(\frac{xy + N}{N}\right)$$

for any $t, x, y \in I$ and $n \in \mathbb{N}$, where $(G_{N,t})_t$ is as in (3.3.16).

In the main result of this subsection, Theorem 3.5.13, we shall derive lower and upper bounds (not depending on $t \in I$) of the supremum

$$\sup_{x \in I, y \in I} |e_{N,n,t}(x, y)|, \quad t \in I,$$

which provide a more refined estimate of the convergence rate involved. First, we obtain a lower bound for the error, which suggests the exact convergence rate of $G_{N,t}\left(s_{n,t}^N \in [0, y]\right)$ to $G_N([0, y])$ as $n \to \infty$ for all $t \in I$.

Theorem 3.5.9 *Whatever $t \in I$ and $n \in \mathbb{N}_+$ we have*

$$\frac{1}{2} V_{N, N(n)}(1) \leq \sup_{y \in I} \left| G_{N,t}\left(s_{n,t}^N \in [0, y]\right) - G_N([0, y]) \right|$$

with $V_{N, N(n)}(t) = \sup_{s \in I} G_{N,t}\left(s_{n,t}^N = s\right)$, where we write $N(n)$ for $(i_1, \ldots, i_n)$ with $i_1 = \cdots = i_n = N$, $n \in \mathbb{N}_+$.

Proof First, the continuity of the function $y \mapsto G_N([0, y])$, $y \in I$, and the equations

$$\lim_{h \searrow 0} G_{N,t}\left(s_{n,t}^N \leq y - h\right) = G_{N,t}\left(s_{n,t}^N < y\right)$$

and

$$\lim_{h \searrow 0} G_{N,t}\left(s_{n,t}^N < y + h\right) = G_{N,t}\left(s_{n,t}^N \leq y\right)$$

imply that

$$\sup_{y \in I} \left| G_{N,t}\left(s_{n,t}^N \leq y\right) - G_N([0, y]) \right| = \sup_{y \in I} \left| G_{N,t}\left(s_{n,t}^N < y\right) - G_N([0, y]) \right|$$

for all $t \in I$ and $n \in \mathbb{N}$. Second, whatever $s \in I$ we have

$$G_{N,t}(s_{n,t}^N = s) = G_{N,t}\left(s_{n,t}^N \leq s\right) - G_N([0, s]) - \left(G_{N,t}\left(s_{n,t}^N < s\right) - G_N([0, s])\right)$$

$$\leq \sup_{y \in I} \left| G_{N,t}\left(s_{n,t}^N \leq y\right) - G_N([0, y]) \right|$$

$$+ \sup_{y \in I} \left| G_{N,t} \left(s_{n,t}^N < y \right) - G_N \left([0, y] \right) \right|$$

$$= 2 \sup_{y \in I} \left| G_{N,t} \left(s_{n,t}^N \leq y \right) - G_N \left([0, y] \right) \right|.$$

Hence

$$\sup_{y \in I} \left| G_{N,t} \left(s_{n,t}^N \in [0, y] \right) - G_N \left([0, y] \right) \right| = \sup_{y \in I} \left| G_{N,t} \left(s_{n,t}^N \leq y \right) - G_N \left([0, y] \right) \right|$$

$$\geq \frac{1}{2} \sup_{s \in I} G_{N,t} \left(s_{n,t}^N = s \right),$$

for all $t \in I$ and $n \in \mathbb{N}$.

By induction with respect to $n \in \mathbb{N}_+$ we get

$$V_N^n f(x) = \sum_{i_1, \ldots, i_n \in \mathbb{N}_N} V_{N, i_1 \ldots i_n}(x) f \left(v_{N, i_n \ldots i_1}(x) \right), \qquad (3.5.33)$$

where $v_{N, i_n \ldots i_1} = v_{N, i_n} \circ \cdots \circ v_{N, i_1}$, $V_{N, i_1 \ldots i_n}(x) = V_{N, i_1}(x) V_{N, i_2}(v_{N, i_1}(x)) \ldots V_{N, i_n}(v_{N, i_{n-1} \ldots i_1}(x))$, $n \geq 2$, and the functions $v_{N, i}$ and $V_{N, i}$, $i \in \mathbb{N}_N$, are defined in (3.5.4).

It follows from (3.3.17) that

$$V_N^n f(t) = E_t(f(s_{n,t}^N)), \; n \in \mathbb{N}, \; f \in B(I), \; t \in I.$$

As $s_{n,t}^N = v_{N, \varepsilon_n \ldots \varepsilon_1}(t), \; t \in I, \; n \in \mathbb{N}_+$, we have

$$V_N^n f(t) = \sum_{i^{(n)} \in (\mathbb{N}_N)^n} G_{N,t} \left((\varepsilon_1, \varepsilon_2, \ldots, \varepsilon_n) = i^{(n)} \right) f \left(v_{N, i_n \ldots i_1}(t) \right) \qquad (3.5.34)$$

for any $n \in \mathbb{N}_+$, $f \in B(I)$, $t \in I$, and $i^{(n)} = (i_1, \ldots, i_n) \in (\mathbb{N}_N)^n$.

Hence, by (3.1.14), (3.5.33), and (3.5.34) we get

$$V_{N, i_1 \ldots i_n}(t) = G_{N,t} \left(I_N \left(i^{(n)} \right) \right) = G_{N,t} \left(s_{n,t}^N = [i_n, \ldots, i_2, i_1 + t]_N \right), \; n \geq 2,$$

$$V_{N, i_1}(t) = G_{N,t} \left(I_N (i_1) \right) = G_{N,t} \left(s_{1,t}^N = \frac{N}{i_1 + t} \right),$$

for all $t \in I$ and $i_1, \ldots, i_n \in \mathbb{N}_N$.

3.5 Gauss-Kuzmin Theorem for N-Continued Fractions

Since as easily seen,

$$\max_{i^{(n)} \in (\mathbb{N}_N)^n} G_{N,t}\left(I_N\left(i^{(n)}\right)\right) = G_{N,t}\left(I_N\left(N(n)\right)\right),$$

where we write $N(n)$ for $i^{(n)} = (i_1, \ldots, i_n)$ with $i_1 = \ldots = i_n = N$, $n \in \mathbb{N}_+$.

In [39] we showed that $I_N\left(i^{(n)}\right)$ is the set of irrationals in the interval with endpoints $\dfrac{p_n}{q_n}$ and $\dfrac{p_n + p_{n-1}}{q_n + q_{n-1}}$. Since

$$\frac{p_n}{q_n} = [i_1, \ldots, i_n]_N = \begin{cases} \dfrac{N}{i_1} & \text{if } n = 1, \\ \dfrac{N}{i_1 + \dfrac{p_{n-1}(i_2, \ldots, i_n)}{q_{n-1}(i_2, \ldots, i_n)}} & \text{if } n \geq 2 \end{cases}$$

and

$$\frac{p_n + p_{n-1}}{q_n + q_{n-1}} = \begin{cases} \dfrac{N}{i_1 + 1} & \text{if } n = 1, \\ [i_1, \ldots, i_{n-1}, i_n + 1]_N & \text{if } n \geq 2, \end{cases}$$

$$= \begin{cases} \dfrac{N}{i_1 + 1} & \text{if } n = 1, \\ \dfrac{N}{i_1 + \dfrac{p_n(i_2, \ldots, i_n, N)}{q_n(i_2, \ldots, i_n, N)}} & \text{if } n \geq 2, \end{cases}$$

we can write

$$V_{N, i_1 \ldots i_n}(t) = \frac{(t + N)N^{n-1}}{q_{n-1}(i_2, \ldots, i_n)(t + i_1) + p_{n-1}(i_2, \ldots, i_n)} \times \frac{1}{q_n(i_2, \ldots, i_n, N)(t + i_1) + p_n(i_2, \ldots, i_n, N)} \quad (3.5.35)$$

for all $i_n \in \mathbb{N}_N$, $n \geq 2$, and $t \in I$.

Also by (3.5.35) we have

$$V_{N,N(n)}(t) = \frac{(t+N)N^{n-1}}{q_{n-1}(\underbrace{N,\ldots,N}_{(n-1)\ times})(t+N) + p_{n-1}(\underbrace{N,\ldots,N}_{(n-1)\ times})}$$

$$\times \frac{1}{q_n(\underbrace{N,\ldots,N,N}_{n\ times})(t+N) + p_n(\underbrace{N,\ldots,N,N}_{n\ times})}.$$

It is easy to see that $V_{N,N(n)}(\cdot)$ is a decreasing function. Therefore

$$\sup_{s \in I} G_{N,t}\left(s_{n,t}^N = s\right) = V_{N,N(n)}(t) \geq V_{N,N(n)}(1)$$

for all $t \in I$. □

Theorem 3.5.10 (The Lower Bound) *Whatever $t \in I$ we have*

$$\frac{1}{2}V_{N,N(n)}(1) \leq \sup_{x \in I, y \in I} \left| G_{N,t}\left(T_N^n \in [0,x], s_{n,t}^N \in [0,y]\right) - \frac{\log\left(\frac{xy+N}{N}\right)}{\log\left(\frac{N+1}{N}\right)} \right|$$

for all $n \in \mathbb{N}_+$.

Proof Whatever $t \in I$ and $n \in \mathbb{N}_+$, by Theorem 3.5.9 we have

$$\sup_{x \in I, y \in I} \left| G_{N,t}\left(T_N^n \in [0,x], s_{n,t}^N \in [0,y]\right) - \frac{1}{\log\left(\frac{N+1}{N}\right)} \log\left(\frac{xy+N}{N}\right) \right|$$

$$\geq \sup_{y \in I} \left| G_{N,t}\left(T_N^n \in I, s_{n,t}^N \in [0,y]\right) - \frac{1}{\log\left(\frac{N+1}{N}\right)} \log\left(\frac{y+N}{N}\right) \right|$$

$$= \sup_{y \in I} \left| G_{N,t}\left(s_{n,t}^N \in [0,y]\right) - G_N([0,y]) \right| \geq \frac{1}{2}V_{N,N(n)}(1).$$

□

Remark 3.5.11 Note that

$$V_{N,N(n)}(1) = \frac{(1+N)N^{n+1}}{\overline{q}_{n+1}\overline{q}_{n+2}}, \quad n \in \mathbb{N}_+,$$

3.5 Gauss-Kuzmin Theorem for N-Continued Fractions

where $\overline{q}_n$ is defined in (3.5.26). Using (3.5.28), it should be noted that Theorem 3.5.10 in connection with the limit

$$\lim_{n \to \infty} \left(\frac{1}{2} V_{N,N(n)}(1) \right)^{1/n} = \frac{2}{N + \sqrt{N^2 + 4N} + 2}$$

leads to an estimate of the order of magnitude of the error $e_{n,t}(x, y)$.

It is known that for the RCF expansion [27] the exact order of magnitude of the supremum $\sup_{x \in I, y \in I} \left| \mu_t \left(\tau^n \in [0, x], s_{n,t} \in [0, y] \right) - \frac{\log(xy + 1)}{\log 2} \right|$, where τ denotes the Gauss map (1.1.2), $\tau^n = [a_{n+1}, a_{n+2}, \ldots]_G$ (= the regular continued fraction with incomplete quotients $a_{n+1}, a_{n+2}, \ldots$), $\mu_t([0, x]) = \frac{(t+1)x}{tx+1}$, $x, t \in I$ and $s_{n+1,t} = 1/(s_{n,t} + a_{n+1})$, $n \geq 0$, $s_{0,t} = t \in I$, is $\mathcal{O}(g^{2n})$ with $g = (\sqrt{5} - 1)/2$, $g^2 = (3 - \sqrt{5})/2 = 0.38196\ldots$. Note that for $N = 1$, $\lim_{n \to \infty} \left(\frac{1}{2} V_{N,N(n)}(1) \right)^{1/n} = g^2$.

In what follows we study the transition operator associated with the RSCC underlying N-continued fraction on the Banach space of complex-valued functions of bounded variation. The characteristic properties of this operator are used to derive an explicit upper bound for $\sup_{x \in I, y \in I} |e_{N,n,t}(x, y)|, t \in I$.

We start by proving the following elementary result.

Theorem 3.5.12 *(The upper bound) Whatever $t \in I$ we have*

$$\sup_{x \in I, y \in I} \left| G_{N,t} \left(T_N^n \in [0, x], s_{n,t}^N \in [0, y] \right) - \frac{\log \left(\frac{xy+N}{N} \right)}{\log \left(\frac{N+1}{N} \right)} \right| \leq \left(\frac{1}{N+1} + \widetilde{K}_N \right)^n$$

for all $n \in \mathbb{N}$.

Proof Let $F_{n,t}^N(y) = G_{N,t}(s_{n,t}^N \leq y)$ and $H_{n,t}^N(y) = F_{n,t}^N(y) - G_N([0, y])$, $t, y \in I$, $n \in \mathbb{N}$. Note that $H_{n,t}^N(0) = 0$. As we have noted V_N is the transition operator of the Markov chain $(s_{n,t}^N)_{n \in \mathbb{N}}$. For any $y \in I$ consider the function f_y defined on I as

$$f_y(t) := \begin{cases} 1 & \text{if } 0 \leq t \leq y, \\ 0 & \text{if } y < t \leq 1. \end{cases}$$

Hence

$$V_N^n f_y(t) = E_t \left(f_y(s_{n,t}^N) \Big| s_{0,t}^N = t \right) = G_{N,t}(s_{n,t}^N \leq y)$$

for all $t, y \in I, n \in \mathbb{N}$. As

$$V_N^\infty f_y = \int_I f_y(t) \mathrm{d}G_N(t) = G_N([0, y]), \quad y \in I.$$

It follows from Corollary 3.4.3 that

$$|H_{n,t}^N(y)| = \left|G_{N,t}(s_{n,t}^N \leq y) - G_N([0, y])\right|$$

$$= |V_N^n f_y(t) - V_N^\infty f_y| \leq \left(\frac{1}{N+1} + \widetilde{K}_N\right)^n \mathrm{var}\, f_y$$

$$= \left(\frac{1}{N+1} + \widetilde{K}_N\right)^n \tag{3.5.36}$$

for all $t, y \in I, n \in \mathbb{N}$. By the very definition of the conditional probability and (3.3.20), for all $t \in I, x, y \in I$ and $n \in \mathbb{N}$, we have

$$G_{N,t}\left(T_N^n \in [0, x], s_{n,t}^N \in [0, y]\right) = G_{N,t}\left(T_N^n \in [0, x] \mid s_{n,t}^N \in [0, y]\right) \cdot G_{N,t}(s_{N,t} \in [0, y])$$

$$= G_{N,t}\left(T_N^n \in [0, x] \mid s_{n,t}^N \in [0, y]\right) \cdot F_{n,t}^N(y)$$

$$= \int_0^y G_{N,t}\left(T_N^n \in [0, x] \mid s_{n,t}^N = z\right) \mathrm{d}F_{n,t}^N(z)$$

$$= \int_0^y \frac{(z+N)x}{zx+N} \mathrm{d}F_{n,t}^N(z) = \int_0^y \frac{(z+N)x}{zx+N} \mathrm{d}G_N(z) + \int_0^y \frac{(z+N)x}{zx+N} \mathrm{d}H_{n,t}^N(z)$$

$$= \frac{1}{\log\left(\frac{N+1}{N}\right)} \log\left(\frac{xy+N}{N}\right) + \frac{(z+N)x}{zx+N} H_{n,t}^N(z)\Big|_0^y$$

$$- \int_0^y \frac{Nx(1-x)}{(zx+N)^2} H_{n,t}^N(z) \mathrm{d}z.$$

Hence, by (3.5.36)

$$\left|G_{N,t}\left(T_N^n \in [0, x], s_{n,t}^N \in [0, y]\right) - \frac{1}{\log\left(\frac{N+1}{N}\right)} \log\left(\frac{xy+N}{N}\right)\right|$$

$$\leq \left(\frac{1}{N+1} + \widetilde{K}_N\right)^n \left(\frac{(y+N)x}{xy+N} + \frac{N(1-x)}{zx+N}\Big|_{z=0}^{z=y}\right)$$

$$= \left(\frac{1}{N+1} + \widetilde{K}_N\right)^n \left(\frac{(y+N)x}{xy+N} - \frac{(1-x)xy}{xy+N}\right)$$

3.5 Gauss-Kuzmin Theorem for N-Continued Fractions

$$= \left(\frac{1}{N+1} + \widetilde{K}_N\right)^n \cdot x \leq \left(\frac{1}{N+1} + \widetilde{K}_N\right)^n,$$

where $\widetilde{K}_N$ is as in (3.4.20), $t, x, y \in I$, $n \in \mathbb{N}$. □

Combining Theorem 3.5.10 with Theorem 3.5.12 we obtain Theorem 3.5.13.

Theorem 3.5.13 *Whatever $t \in I$ we have*

$$\frac{1}{2} V_{N,N(n)}(1) \leq \sup_{x \in I, y \in I} \left| G_{N,t}\left(T_N^n \in [0,x], s_{n,t}^N \in [0,y]\right) - \frac{\log\left(\frac{xy+N}{N}\right)}{\log\left(\frac{N+1}{N}\right)} \right|$$

$$\leq \left(\frac{1}{N+1} + \widetilde{K}_N\right)^n$$

for all $n \in \mathbb{N}_+$.

Remark 3.5.14 Theorem 3.5.13 implies that the convergence rate is $\mathcal{O}(\gamma_N^n)$, with

$$\frac{2}{N + \sqrt{N^2 + 4N} + 2} \leq \gamma_N \leq \frac{1}{N+1} + \widetilde{K}_N.$$

For example, we have (Table 3.1):

Table 3.1 Lower and upper bounds for some $N \geq 1$

$N = 1$	$g^2 = 0.381966 \leq \gamma_N \leq 0.843146$
$N = 2$	$0.267949192 \leq \gamma_N \leq 0.535374333\ldots$
$N = 5$	$0.145898033 \leq \gamma_N \leq 0.257764367\ldots$
$N = 10$	$0.083920216 \leq \gamma_N \leq 0.138555191\ldots$
$N = 100$	$0.009804864 \leq \gamma_N \leq 0.01487615$
$N = 1000$	$0.000998004 \leq \gamma_N \leq 0.001498751$
$N = 10000$	$0.00009998 \leq \gamma_N \leq 0.00009999$

Generalized Rényi Continued Fractions

4

In this chapter we introduce and study in detail a special class of backward continued fractions that represents a generalization of Rényi continued fractions.

4.1 Introduction

In [22], Gröchenig and Haas investigated u-backward continued fractions associated with the one-parameter family of interval maps of the form $T_u(x) := \frac{1}{u(1-x)} - \lfloor \frac{1}{u(1-x)} \rfloor$, where $u > 0$, $x \in [0, 1)$, and $\lfloor \cdot \rfloor$ denotes the integer part. As the parameter varies, the maps T_u exhibit a curious dynamical behavior. It was known that varying u in the interval $(0, 4)$ gives a viable theory of u-backward continued fractions, which fails when $u \geq 4$. So, for a given $u \in (0, 4)$, each irrational $x \in [0, 1)$ can be uniquely represented as an infinite continued fraction

$$x = 1 - \cfrac{1}{un_1 - \cfrac{1}{n_2 - \cfrac{1}{un_3 - \cfrac{\cdot}{\cdot\cdot} - \cfrac{1}{s_k n_k - \cdot\cdot\cdot}}}} =: [r_1, r_2, r_3, \ldots]_u, \qquad (4.1.1)$$

where the integers $n_i = 1 + r_i \geq 2$ and the "coefficient" s_j of n_j alternates between $s_j = 1$ for even j and $s_j = u$ for odd j. The case $u = 1$ was studied by Rényi [53] and provides an alternate approach to continued fractions and rational approximation. The graph of T_1 can be obtained from the graph of the regular (Gauss) continued fraction transformation $\tau(x) := \frac{1}{x} - \lfloor \frac{1}{x} \rfloor$ by reflecting about the line $x = \frac{1}{2}$ (Fig. 4.1).

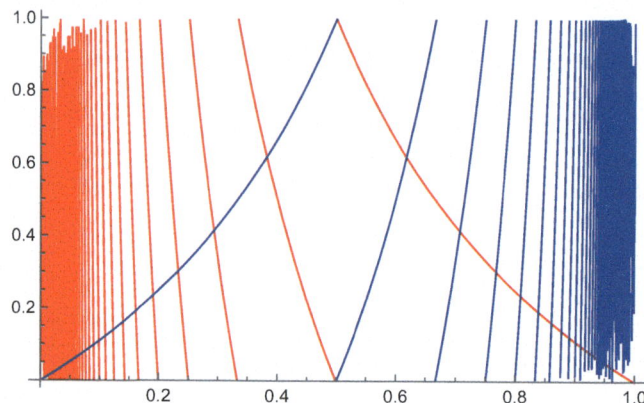

Fig. 4.1 Graphs of Gauss and Rényi transformations

From the dynamical systems point of view, interest in the Rényi map steams from two facts. First, it has a parabolic fixed point, $T_1(0) = 0$ and $T_1'(0) = 1$, and therefore it exhibits non-uniformly hyperbolic behavior. Secondly, it has infinite entropy. Therefore, the Rényi map exhibits two sources of complications that are of independent interest and provides an example where we have the opportunity to study how they interact. These features have attracted attention, and interesting results have been obtained describing the ergodic theory of this map. In particular Adler and Flatto [1] showed that there exists an infinite invariant measure absolutely continuous with respect to the Lebesgue measure. There is no finite one.

The main purpose of Gröchenig and Haas ([22, 23]) was to find an explicit form for the absolutely continuous invariant measure for T_u similar to that of the Gauss measure $\frac{dx}{x+1}$ for τ and Rényi's measure $\frac{dx}{x}$ for T_1. While the Gauss measure is finite, the Rényi's measure is infinite. They showed that the invariant measure for T_u is finite if and only if $0 < u < 4$ and $u = u_q \neq 4\cos^2\frac{\pi}{q}$, $q = 3, 4, \ldots$. They also identified the invariant probability measure for T_u corresponding to the values $u = 1/N$ for positive integers $N \geq 2$. This is a finite measure. In this particular case we will call the continued fraction in (4.1.1) a *Rényi-type continued fraction*. We will indicate $T_{\frac{1}{N}}$ by R_N, and we will call it a *Rényi-type (continued fraction) transformation*.

Our goal in this chapter is to start an approach to the metrical theory of Rényi-type continued fraction expansions via dependence with complete connections. Using the natural extensions for the Rényi-type transformations, we give an infinite-order-chain representation of the sequence of the incomplete quotients of these expansions. We then show that the associated random systems with complete connections are with contraction and their transition operators are regular with respect to the Banach space of Lipschitz functions. Further this leads, in Sect. 4.6, to a solution of the Gauss-Kuzmin problem for these continued fractions which allows us to obtain a convergence rate result.

4.2 Rényi-Type Continued Fraction Expansion as Dynamical System

For a fixed integer $N \geq 2$, we define the Rényi-type continued fraction transformation $R_N : I \to I$, $I := [0, 1]$, by

$$R_N(x) := \begin{cases} \dfrac{N}{1-x} - \left\lfloor \dfrac{N}{1-x} \right\rfloor & \text{if } x \neq 1 \\ 0 & \text{if } x = 1. \end{cases} \quad (4.2.1)$$

To this transformation, we associate the digits $r_n(x)$, $n \in \mathbb{N}_+ := \{1, 2, 3, \ldots\}$, which are defined by

$$r_n = r_n(x) := r_1\left(R_N^{n-1}(x)\right), \quad n \geq 2, \quad (4.2.2)$$

with $R_N^0(x) = x$ and

$$r_1 = r_1(x) := \begin{cases} \left\lfloor \dfrac{N}{1-x} \right\rfloor & \text{if } x \neq 1, \\ \infty & \text{if } x = 1. \end{cases} \quad (4.2.3)$$

Here and hereafter we omit the index N whenever it is clear from context.

We observe that $r_n \in \mathbb{N}_N := \{N, N+1, \ldots\}$ for any $n \in \mathbb{N}_+$. Using these digits, we can write

$$R_N(x) = \frac{N}{1-x} - r_1(x) \quad (4.2.4)$$

and

$$x = 1 - \cfrac{N}{1 + r_1 - \cfrac{N}{1 + r_2 - \cfrac{N}{1 + r_3 - \cfrac{\cdot}{\cdot \cdot}}}} =: [r_1, r_2, r_3, \ldots]_R. \quad (4.2.5)$$

The Rényi-type continued fraction in (4.2.5) can be viewed as a measure preserving dynamical system $(I, \mathcal{B}_I, R_N, \rho_N)$, where $\mathcal{B}_I$ denotes the σ-algebra of all Borel subsets of I, and

$$\rho_N(A) := \frac{1}{\log\left(\frac{N}{N-1}\right)} \int_A \frac{dx}{x + N - 1}, \quad A \in \mathcal{B}_{[0,1]} \quad (4.2.6)$$

is the invariant probability measure under R_N [22].

Let $\frac{p_n}{q_n} := \frac{p_n(x)}{q_n(x)}$ denote the partial fractions obtained by applying the transformation R_N n times, so

$$\frac{p_n}{q_n} = 1 - \cfrac{N}{1 + r_1 - \cfrac{N}{1 + r_2 - \cdots - \cfrac{N}{1 + r_n}}}. \qquad (4.2.7)$$

For the functions p_n and q_n we obtain the following recurrence relations:

$$p_0 = 1, \; p_1 = 1 + r_1 - N, \quad p_n := (1 + r_n)p_{n-1} - Np_{n-2}, n \geq 2, \qquad (4.2.8)$$

$$q_0 = 1, \; q_1 = 1 + r_1, \quad\quad\quad q_n := (1 + r_n)q_{n-1} - Nq_{n-2}, n \geq 2. \qquad (4.2.9)$$

Using these recurrences, induction easily gives that

$$p_{n-1}q_n - p_n q_{n-1} = N^n, \quad n \in \mathbb{N}_+ \qquad (4.2.10)$$

and

$$x = \frac{p_n - (1 - R_N^n)p_{n-1}}{q_n - (1 - R_N^n)q_{n-1}}, \quad n \in \mathbb{N}_+. \qquad (4.2.11)$$

We obtain

$$\left| x - \frac{p_n}{q_n} \right| \leq \frac{N^n}{q_n(q_n - q_{n-1})}, \quad n \in \mathbb{N}_+. \qquad (4.2.12)$$

4.3 The Probabilistic Structure of $(r_n)_{n \in \mathbb{N}_+}$ under the Lebesgue Measure

We start by defining the n-th order cylinder associated with the digits $(r_n)_{n \in \mathbb{N}_+}$ of the Rényi-type continued fraction (4.2.5). An n-block $(r_1, r_2, \ldots, r_n)$ is said to be *admissible* for the expansion in (4.2.5) if there exists $x \in [0, 1)$ such that $r_i(x) = r_i$ for all $1 \leq i \leq n$. If $(r_1, r_2, \ldots, r_n)$ is an admissible sequence, we call the set

$$I_R(r_1, r_2, \ldots, r_n) = \{x \in I : r_1(x) = r_1, r_2(x) = r_2, \ldots, r_n(x) = r_n\}, \qquad (4.3.1)$$

the n-th order cylinder. As we mentioned above, $(r_1, r_2, \ldots, r_n) \in (\mathbb{N}_N)^n$. For example, for any $r_1 = i \in \mathbb{N}_N$ we have

4.3 The Probabilistic Structure of $(r_n)_{n\in\mathbb{N}_+}$ under the Lebesgue Measure

$$I_R(r_1) = \{x \in I : r_1(x) = r_1\} = \left[1 - \frac{N}{i}, 1 - \frac{N}{i+1}\right). \quad (4.3.2)$$

By induction, it can be shown that

$$I_R(r_1, r_2, \ldots, r_n) = \left[\frac{p_n - p_{n-1}}{q_n - q_{n-1}}, \frac{p_n}{q_n}\right), \quad (4.3.3)$$

for all $n \geq 1$. Hence, using (4.2.10) we have

$$\lambda(I_R(r_1, r_2, \ldots, r_n)) = \frac{N^n}{q_n(q_n - q_{n-1})}, \quad (4.3.4)$$

where λ is the Lebesgue measure and $\{q_n\}$ is as in (4.2.9).

To derive the so-called Brodén-Borel-Lévy formula [28, 29], let us define $(s_{N,n})_{n\in\mathbb{N}}$ by

$$s_{N,0} := 1, \quad s_{N,n} := 1 - N\frac{q_{n-1}}{q_n}. \quad (4.3.5)$$

From (4.2.9), $s_{N,n} = 1 - N/(r_n + s_{n-1})$. Hence,

$$s_{N,n} = 1 - \cfrac{N}{1 + r_n - \cfrac{N}{1 + r_{n-1} - \cfrac{\ddots}{ - \cfrac{N}{1+r_1}}}} = [r_n, r_{n-1}, \ldots, r_1]_R. \quad (4.3.6)$$

Proposition 4.3.1 (Brodén-Borel-Lévy-Type Formula [42]) *Let λ denote the Lebesgue measure on I. For any $n \in \mathbb{N}_+$, the conditional probability $\lambda(R_N^n < x | r_1, \ldots, r_n)$ is given by*

$$\lambda(R_N^n < x | r_1, \ldots, r_n) = \frac{Nx}{N - (1-x)(1 - s_{N,n})}, \quad x \in I, \quad (4.3.7)$$

where $\{s_{N,n}\}$ is as in (4.3.5) and $r_1, \ldots, r_n$ are as in (4.2.2) and (4.2.3).

Proof By definition, we have

$$\lambda\left(R_N^n < x | r_1, \ldots, r_n\right) = \frac{\lambda\left(\left(R_N^n < x\right) \cap I_R(r_1, \ldots, r_n)\right)}{\lambda(I_R(r_1, \ldots, r_n))}$$

for any $n \in \mathbb{N}_+$ and $x \in I$. From (4.2.11) and (4.3.3) we have

$$\lambda\left(\left(R_N^n < x\right) \cap I_R(r_1, \ldots, r_n)\right) = \left|\frac{p_n + (x-1)p_{n-1}}{q_n + (x-1)q_{n-1}} - \frac{p_n - p_{n-1}}{q_n - q_{n-1}}\right|$$

$$= \frac{N^n x}{(q_n - q_{n-1})(q_n + (x-1)q_{n-1})}.$$

From this and (4.3.4), we have that

$$\lambda\left(R_N^n < x | r_1, \ldots, r_n\right) = \frac{\lambda\left(\left(R_N^n < x\right) \cap I_R(r_1, \ldots, r_n)\right)}{\lambda\left(I_R(r_1, \ldots, r_n)\right)}$$

$$= \frac{xq_n}{q_n + (x-1)q_{n-1}} = \frac{Nx}{N - (1-x)(1-s_{N,n})}$$

for any $n \in \mathbb{N}_+$ and $x \in I$. □

The Brodén-Borel-Lévy-type formula allows us to determine the probabilistic structure of the digits $(r_n)_{n \in \mathbb{N}_+}$ under λ.

Proposition 4.3.2 ([42]) *For any $i \geq N$ and $n \in \mathbb{N}_+$, we have*

$$\lambda(r_1 = i) = \frac{N}{i(i+1)}, \quad \lambda(r_{n+1} = i | r_1, \ldots, r_n) = P_{N,i}(s_{N,n}), \quad (4.3.8)$$

where $\{s_{N,n}\}$ is as in (4.3.5), and

$$P_{N,i}(x) := \frac{x + N - 1}{(x+i)(x+i-1)}. \quad (4.3.9)$$

Proof From (4.3.2), the case $\lambda(r_1 = i)$ holds. By using (4.2.2) and (4.3.7), we have

$$\lambda(r_{n+1} = i | r_1, \ldots, r_n) = \lambda\left(R_N^n \in \left[1 - \frac{N}{i}, 1 - \frac{N}{i+1}\right) | r_1, \ldots, r_n\right)$$

$$= \frac{N\left(1 - \frac{N}{i+1}\right)}{N - \frac{N}{i+1}(1 - s_{N,n})} - \frac{N\left(1 - \frac{N}{i}\right)}{N - \frac{N}{i}(1 - s_{N,n})} = P_{N,i}(s_{N,n}).$$

□

It is easy to check that $\sum_{i \geq N} P_{N,i}(x) = 1$ for any $x \in I$.

Remark 4.3.3 Proposition 4.3.2 is the starting point of an approach to the metrical theory of Rényi-type continued fraction expansions via dependence with complete connections (see [28], Section 5.2). We apply this method in Sect. 4.6 to obtain a solution of Gauss-Kuzmin-type problem for Rényi-type continued fraction expansions.

Corollary 4.3.4 ([42]) *The sequence* $(s_{N,n})_{n \in \mathbb{N}_+}$ *with* $s_{N,0} = 1$ *is a homogeneous I-valued Markov chain on* $(I, \mathcal{B}_I, \lambda)$ *with the following transition mechanism: From state $s \in I$ the only possible one-step transitions are those of states $1 - N/(s+i)$, $i \geq N$, with corresponding probabilities $P_{N,i}(s)$, $i \geq N$.*

4.4 Natural Extension and Extended Random Variables

Fix an integer $N \geq 2$. In this section, we introduce the natural extension $\overline{R}_N$ of R_N in (4.2.1) and its extended random variables [29].

4.4.1 Natural Extension of R_N

Let $(I, \mathcal{B}_I, R_N)$ be as in Sect. 4.2. Define $(u_{N,i})_{i \geq N}$ by

$$u_{N,i} : I \to I; \quad u_{N,i}(x) := 1 - \frac{N}{x+i}, \quad x \in I. \tag{4.4.1}$$

For each $i \geq N$, $u_{N,i}$ is a right inverse of R_N, that is,

$$\left(R_N \circ u_{N,i}\right)(x) = x, \quad \text{for any } x \in I. \tag{4.4.2}$$

Furthermore, if $r_1(x) = i$, then $\left(u_{N,i} \circ R_N\right)(x) = x$ where r_1 is as in (4.2.3).

Definition 4.4.1 The natural extension $\left(I^2, \mathcal{B}_I^2, \overline{R}_N\right)$ of $(I, \mathcal{B}_I, R_N)$ is the transformation $\overline{R}_N$ of the square space $\left(I^2, \mathcal{B}_I^2\right) := (I, \mathcal{B}_I) \times (I, \mathcal{B}_I)$ defined as follows [49]:

$$\overline{R}_N : I^2 \to I^2; \quad \overline{R}_N(x, y) := \left(R_N(x), u_{N, r_1(x)}(y)\right), \quad (x, y) \in I^2. \tag{4.4.3}$$

From (4.4.2), we see that $\overline{R}_N$ is bijective on I^2 with the inverse

$$(\overline{R}_N)^{-1}(x, y) = (u_{N, r_1(y)}(x), R_N(y)), \quad (x, y) \in I^2. \tag{4.4.4}$$

Iterations of (4.4.3) and (4.4.4) are given as follows for each $n \geq 2$:

$$\left(\overline{R}_N\right)^n (x, y) = \left(R_N^n(x), [r_n(x), r_{n-1}(x), \ldots, r_2(x), r_1(x) + y - 1]_R\right),$$
$$\left(\overline{R}_N\right)^{-n} (x, y) = \left([r_n(y), r_{n-1}(y), \ldots, r_2(y), r_1(y) + x - 1]_R, R_N^n(y)\right).$$

For ρ_N in (4.2.6), we define its *extended measure* $\overline{\rho}_N$ on $\left(I^2, \mathcal{B}_I^2\right)$ as

$$\overline{\rho}_N(B) := \frac{1}{\log\left(\frac{N}{N-1}\right)} \iint_B \frac{N\,dx\,dy}{\{N-(1-x)(1-y)\}^2}, \quad B \in \mathcal{B}_I^2. \tag{4.4.5}$$

Then $\overline{\rho}_N(A \times I) = \overline{\rho}_N(I \times A) = \rho_N(A)$ for any $A \in \mathcal{B}_I$. The measure $\overline{\rho}_N$ is preserved by $\overline{R}_N$, i.e., $\overline{\rho}_N((\overline{R}_N)^{-1}(B)) = \overline{\rho}_N(B)$ for any $B \in \mathcal{B}_I^2$.

4.4.2 Extended Random Variables

With respect to $\overline{R}_N$ in (4.4.3), define *extended incomplete quotients* $\overline{r}_l(x,y)$, $l \in \mathbb{Z} := \{\ldots, -2, -1, 0, 1, 2, \ldots\}$ at $(x,y) \in I^2$ by

$$\overline{r}_{l+1}(x,y) := r_1\left((\overline{R}_N)^l(x,y)\right), \quad l \in \mathbb{Z}, \tag{4.4.6}$$

with $\overline{r}_1(x,y) = r_1(x)$, $x, y \in [0,1]$.

Remark 4.4.2

(i) Note that $\overline{r}_l(x,y)$ in (4.4.6) is also well-defined for $l \leq 0$ because $\overline{R}_N$ is invertible. By (4.4.3) and (4.4.4) we have

$$\overline{r}_n(x,y) = r_n(x), \quad \overline{r}_0(x,y) = r_1(y), \quad \overline{r}_{-n}(x,y) = r_{n+1}(y), \tag{4.4.7}$$

for any $n \in \mathbb{N}_+$ and $(x,y) \in I^2$.

(ii) Since the measure $\overline{\rho}_N$ is preserved by $\overline{R}_N$, the doubly infinite sequence $(\overline{r}_l(x,y))_{l \in \mathbb{Z}}$ is strictly stationary (i.e., its distribution is invariant under a shift of the indices) under $\overline{\rho}_N$. It is indeed a doubly infinite version of $(r_n)_{n \in \mathbb{N}_+}$ under ρ_N.

Theorem 4.4.3 ([42]) *Fix* $(x,y) \in I^2$ *and let* $\overline{r}_l := \overline{r}_l(x,y)$ *for* $l \in \mathbb{Z}$. *Set* $r := [\overline{r}_0, \overline{r}_{-1}, \ldots]_R$. *Then the following holds for any* $x \in I$:

$$\overline{\rho}_N([0,x] \times I \mid \overline{r}_0, \overline{r}_{-1}, \ldots) = \frac{Nx}{N-(1-x)(1-r)} \quad \overline{\rho}_N\text{-a.s.} \tag{4.4.8}$$

The stochastic property of $(\overline{r}_l)_{l \in \mathbb{Z}}$ under $\overline{\rho}_N$ is as follows.

Corollary 4.4.4 ([42]) *For any* $i \geq N$, *we have*

$$\overline{\rho}_N(\overline{r}_1 = i \mid \overline{r}_0, \overline{r}_{-1}, \ldots) = P_{N,i}(r) \quad \overline{\rho}_N\text{-a.s.,} \tag{4.4.9}$$

where $r := [\overline{r}_0, \overline{r}_{-1}, \ldots]_R$ *and* $P_{N,i}$ *is as in (4.3.9).*

4.4 Natural Extension and Extended Random Variables

Remark 4.4.5 ([42]) The strict stationarity of $(\overline{r}_l)_{l \in \mathbb{Z}}$, under $\overline{\rho}_N$, implies that

$$\overline{\rho}_N(\overline{r}_{l+1} = i \mid \overline{r}_l, \overline{r}_{l-1}, \ldots) = P_{N,i}(r) \quad \overline{\rho}_N\text{-a.s.} \tag{4.4.10}$$

for any $i \geq N$ and $l \in \mathbb{Z}$, where $r := [\overline{r}_l, \overline{r}_{l-1}, \ldots]_R$. Equation (4.4.10) emphasizes that $(\overline{r}_l)_{l \in \mathbb{Z}}$ is a chain of infinite order in the theory of dependence with complete connections [28].

Define extended random variables $(\overline{s}_{N,l})_{l \in \mathbb{Z}}$ by $\overline{s}_{N,l} := [\overline{r}_l, \overline{r}_{l-1}, \ldots]_R$, $l \in \mathbb{Z}$. Clearly, $\overline{s}_{N,l} = \overline{s}_{N,0} \circ (\overline{R}_N)^l$, $l \in \mathbb{Z}$. It follows from Corollary 4.4.4 that $(\overline{s}_{N,l})_{l \in \mathbb{Z}}$ is a strictly stationary $[0, 1)$-valued Markov process on $(I^2, \mathcal{B}_I^2, \overline{\rho}_N)$ with the following transition mechanism. From state $\overline{s} \in I$ the possible transitions are to any state $1 - N/(\overline{s} + i)$ with corresponding transition probability $P_{N,i}(\overline{s})$, $i \geq N$. Clearly, for any $l \in \mathbb{Z}$ we have

$$\overline{\rho}_N(\overline{s}_{N,l} < x) = \overline{\rho}_N(I \times [0, x)) = \rho_N([0, x)), \quad x \in I. \tag{4.4.11}$$

Motivated by Theorem 4.4.3, we shall consider the one-parameter family $\{\rho_{N,a} : a \in I\}$ of (conditional) probability measures on $(I, \mathcal{B}_I)$ defined by their distribution functions

$$\rho_{N,a}([0, x]) := \frac{Nx}{N - (1-x)(1-a)}, \quad x, a \in I. \tag{4.4.12}$$

Note that $\rho_{N,1} = \lambda$. For any $a \in I$ put

$$s_{N,0}^a := a, \quad s_{N,n}^a := 1 - \frac{N}{r_n + s_{N,n-1}^a}, \quad n \in \mathbb{N}_+. \tag{4.4.13}$$

Remark 4.4.6 ([42]) It follows from the properties just described for the process $(\overline{s}_{N,l})_{l \in \mathbb{Z}}$ that the sequence $(s_{N,n}^a)_{n \in \mathbb{N}_+}$ is an I-valued Markov chain on $(I, \mathcal{B}_I, \rho_{N,a})$, which starts at $s_{N,0}^a = a$ and has the following transition mechanism: From state $s \in I$ the possible transitions are to any state $1 - N/(s+i)$ with corresponding transition probability $P_{N,i}(s)$, $i \geq N$.

The transition operator of $(s_{N,n}^a)_{n \in \mathbb{N}_+}$ takes $f \in B(I)$ into the function defined by

$$E_{\rho_{N,a}}\left(f(s_{N,n+1}^a) \mid s_{N,n}^a = s\right) = \sum_{i \geq N} P_{N,i}(s) f\left(u_{N,i}(s)\right) = U_N f(s) \tag{4.4.14}$$

for any $s \in I$, where $E_{\rho_{N,a}}$ stands for the mean value operator with respect to the probability measure $\rho_{N,a}$, whatever $a \in I$, and U_N is the Perron-Frobenius operator of $(I, \mathcal{B}_I, \rho_N, R_N)$ defined as in (4.5.2).

Note that for any $a \in I$ and $n \in \mathbb{N}_+$ we have

$$\rho_{N,a}(A|r_1, \ldots, r_n) = \rho_{N, s_{N,n}^a}\left(R_N^n(A)\right),$$

whatever the set A belonging to the σ-algebra generated by the random variables r_{n+1}, $r_{n+2}\ldots$, that is, $\sigma(r_{n+1}, r_{n+2}, \ldots) = R_N^{-n}(\mathcal{B}_I)$. In particular, it follows that the Brodén-Borel-Lévy formula [29, 42] holds under $\rho_{N,a}$ for any $a \in I$, that is,

$$\rho_{N,a}(R_N^{-n}([0, x])|r_1, \ldots, r_n) = \frac{Nx}{N - (1-x)\left(1 - s_{N,n}^a\right)}, \quad x \in I, n \in \mathbb{N}_+.$$
(4.4.15)

4.5 Perron-Frobenius Operators

Let $(I, \mathcal{B}_I, R_N, \rho_N)$ be as in Sect. 4.2. In this section, we derive its Perron-Frobenius operator.

Let μ be a probability measure on $(I, \mathcal{B}_I)$ such that $\mu\left(R_N^{-1}(A)\right) = 0$ whenever $\mu(A) = 0$ for $A \in \mathcal{B}_I$. If R_N is μ-preserving, this condition is satisfied. Let $L^1(I, \mu) := \{f : I \to \mathbb{C} : \int_0^1 |f| d\mu < \infty\}$. The *Perron-Frobenius operator* of $(I, \mathcal{B}_I, R_N, \mu)$ is the linear operator on the Banach space $L^1(I, \mu)$ satisfying:

$$\int_A P_\mu f \, d\mu = \int_{R_N^{-1}(A)} f \, d\mu \quad \text{for all } A \in \mathcal{B}_I, f \in L^1(I, \mu).$$
(4.5.1)

Proposition 4.5.1 ([42]) *Let $(I, \mathcal{B}_I, R_N, \rho_N)$ be as in Sect. 4.2:*

(i) *The Perron-Frobenius operator $U_N := P_{\rho_N}$ of R_N under the invariant measure ρ_N on $\mathcal{B}_I$ is given a.e. in I by the equation*

$$U_N f(x) = \sum_{i \geq N} P_{N,i}(x) f\left(u_{N,i}(x)\right), \quad f \in L^1(I, \rho_N),$$
(4.5.2)

where $P_{N,i}$ and $u_{N,i}$ are as in (4.3.9) and (4.4.1), respectively.

(ii) *Let μ be a probability measure on $(I, \mathcal{B}_I)$ such that $\mu \ll \lambda$ and let $h := d\mu/d\lambda$ a.e. in $[0, 1]$. Then for any $n \in \mathbb{N}_+$ and $A \in \mathcal{B}_I$, we have*

$$\mu\left(R_N^{-n}(A)\right) = \int_A U_N^n f_N(x) d\rho_N(x),$$
(4.5.3)

where $f_N(x) := \left(\log\left(\frac{N}{N-1}\right)\right)(x + N - 1)h(x)$ for $x \in I$.

4.5 Perron-Frobenius Operators

Proof

(i) Let $R_{N,i}$ denote the restriction of R_N to the subinterval $I_R(i) := \left(1 - \frac{N}{i}, 1 - \frac{N}{i+1}\right]$, $i \geq N$, that is,

$$R_{N,i}(x) = \frac{N}{1-x} - i, \quad x \in I_R(i). \tag{4.5.4}$$

Let $C(A) := R_N^{-1}(A)$ and $C_i(A) := R_{N,i}^{-1}(A)$ for $A \in \mathcal{B}_I$. Since $C(A) = \bigcup_i C_i(A)$ and $C_i \cap C_j$ is a null set when $i \neq j$, we have

$$\int_{C(A)} f \, d\rho_N = \sum_{i \geq N} \int_{C_i(A)} f \, d\rho_N, \quad f \in L^1(I, \rho_N), \, A \in \mathcal{B}_I. \tag{4.5.5}$$

For any $i \geq N$, by the change of variable $x = R_{N,i}^{-1}(y) = 1 - \frac{N}{y+i}$, we obtain

$$\int_{C_i(A)} f(x) \rho_N(dx) = \left(\log\left(\frac{N}{N-1}\right)\right)^{-1} \int_{C_i(A)} \frac{f(x)}{x + N - 1} dx$$

$$= \left(\log\left(\frac{N}{N-1}\right)\right)^{-1} \int_A \frac{1}{(y+i)(y+i-1)} f\left(u_{N,i}(y)\right) dy$$

$$= \int_A P_{N,i}(y) f\left(u_{N,i}(y)\right) \rho_N(dy). \tag{4.5.6}$$

Now, (4.5.2) follows from (4.5.5) and (4.5.6).

(ii) We will use mathematical induction. For $n = 0$, Eq. (4.5.3) holds by the definitions of f and h. Assume that (4.5.3) holds for some $n \in \mathbb{N}$. Then

$$\mu\left(R_N^{-n-1}(A)\right) = \mu\left(R_N^{-n}\left(R_N^{-1}(A)\right)\right) = \int_{C(A)} U_N^n f(x) \rho_N(dx), \tag{4.5.7}$$

and by definition, we have

$$\int_{C(A)} U_N^n f(x) \rho_N(dx) = \int_A U_N^{n+1} f(x) \rho_N(dx). \tag{4.5.8}$$

Therefore,

$$\mu\left(R_N^{-(n+1)}(A)\right) = \int_A U_N^{n+1} f(x) \rho_N(dx), \tag{4.5.9}$$

which ends the proof. $\square$

Next, we calculate the variation of the Perron-Frobenius operator U_N.

Proposition 4.5.2 ([68]) *For any $f \in BV(I)$ we have*

$$\operatorname{var} U_N f \leq \frac{1}{N} \cdot \operatorname{var} f + K_N \cdot |f|,$$

where

$$K_N := \frac{2}{2N - 1 + 2\sqrt{N(N-1)}}. \tag{4.5.10}$$

Proof Recall that $P_{N,i}(x) = \dfrac{i+1-N}{x+i} - \dfrac{i-N}{x+i-1}, i \geq N$. We have

$$\left(P_{N,i}(x)\right)' = \frac{i-N}{(x+i-1)^2} - \frac{i+1-N}{(x+i)^2} = \frac{L(N,x)}{(x+i-1)^2(x+i)^2}$$

with $L(N, x) = -x^2 + 2x(1-N) + i^2 + i(1-2N) + N - 1$, for every $i \geq N$.

If $N \leq i \leq 2N-2$, then $L(N, x) < 0$ for all $x \in I$, i.e., $\left(P_{N,i}(x)\right)' < 0, x \in I$. Hence, the functions $P_{N,i}$ are decreasing on I.

If $i = 2N - 1$, then $L(N, x) > 0$ for all $x \in \left[0, 1 - N + \sqrt{N(N-1)}\right]$, and $L(N, x) < 0$ for all $x \in \left(1 - N + \sqrt{N(N-1)}, 1\right]$. Hence $P_{N,2N-1}$ is increasing on $\left[0, 1 - N + \sqrt{N(N-1)}\right]$ and decreasing on $\left(1 - N + \sqrt{N(N-1)}, 1\right]$.

If $i \geq 2N$, then $L(N, x) > 0$ for all $x \in I$, i.e., $\left(P_{N,i}(x)\right)' > 0, x \in I$. Hence, the functions $P_{N,i}$ are increasing on I.

Hence

$$\operatorname{var} P_{N,i} = \begin{cases} P_{N,i}(0) - P_{N,i}(1) & \text{if } N \leq i \leq 2N-2 \\ 2P_{N,i}(1 - N + \sqrt{N(N-1)}) - P_{N,i}(0) - P_N^i(1) & \text{if } i = 2N-1 \\ P_{N,i}(1) - P_{N,i}(0) & \text{if } i \geq 2N \end{cases}$$

and

$$|P_{N,i}| = \sup_{x \in I} P_{N,i}(x) = \begin{cases} P_{N,i}(0) & \text{if } N \leq i \leq 2N-2 \\ P_{N,i}(1 - N + \sqrt{N(N-1)}), & \text{if } i = 2N-1 \\ P_{N,i}(1) & \text{if } i \geq 2N. \end{cases}$$

Thus

$$\sup_{i \geq N} |P_{N,i}| = \max\left\{P_{N,N}(0), P_{N,2N-1}(1 - N + \sqrt{N(N-1)}), P_{N,2N}(1)\right\}$$

4.5 Perron-Frobenius Operators

$$= \max\left\{\frac{1}{N}, \frac{1}{(\sqrt{N}+\sqrt{N-1})^2}, \frac{1}{2(1+2N)}\right\} = \frac{1}{N}.$$

Also,

$$\sum_{i \geq N} \text{var } P_{N,i} = \sum_{N \leq i \leq 2N-2} \left(P_{N,i}(0) - P_{N,i}(1)\right) + \text{var } P_{N,2N-1} + \sum_{i \geq 2N} \left(P_{N,i}(1) - P_{N,i}(0)\right)$$

$$= \frac{1}{2(2N-1)} + \frac{2}{2N-1+2\sqrt{N(N-1)}} - \frac{1}{2N-1} + \frac{1}{2(2N-1)}$$

$$= \frac{2}{2N-1+2\sqrt{N(N-1)}}.$$

Taking into account that

$$\sum_{i \geq N} \text{var}(f \circ u_{N,i}) = \sum_{i \geq N} \text{var}_{\left[1-\frac{N}{i}, 1-\frac{N}{i+1}\right]} f = \text{var } f,$$

we have

$$\text{var } U_N f = \text{var } \sum_{i \geq N} P_{N,i} \cdot (f \circ u_{N,i}) \leq \sum_{i \geq N} \text{var}\left(P_{N,i} \cdot (f \circ u_{N,i})\right)$$

$$\leq \sum_{i \geq N} |P_{N,i}| \text{var}(f \circ u_{N,i})$$

$$+ \sum_{i \geq N} |f \circ u_{N,i}| \text{var } P_{N,i} \leq \left(\sup_{i \geq N} |P_{N,i}|\right) \sum_{i \geq N} \text{var}(f \circ u_{N,i}) + |f| \sum_{i \geq N} \text{var } P_{N,i}$$

$$\leq \frac{1}{N} \cdot \text{var } f + K_N \cdot |f|,$$

where the constant K_N is as in (4.5.10). $\square$

If $f \in B(I)$, define the linear functional U_N^∞ by

$$U_N^\infty : B(I) \to \mathbb{C}; \quad U_N^\infty f := \int_I f(x) \, d\rho_N(x). \tag{4.5.11}$$

Then, by the invariance of R_N under ρ_N, we have

$$U_N^\infty U_N^n f = U_N^\infty f \quad \text{for any } n \in \mathbb{N}_+. \tag{4.5.12}$$

Corollary 4.5.3 ([68]) *For any $f \in BV(I)$ and for all $n \in \mathbb{N}$ we have*

$$\operatorname{var} U_N^n f \leq \left(\frac{1}{N} + K_N\right)^n \cdot \operatorname{var} f, \qquad (4.5.13)$$

$$\left|U_N^n f - U_N^\infty f\right| \leq \left(\frac{1}{N} + K_N\right)^n \cdot \operatorname{var} f. \qquad (4.5.14)$$

Proof Note that for any $f \in BV(I)$ and $u \in I$, since $\int_I d\rho_N(x) = 1$, we have

$$|f(u)| - \left|\int_I f(x) d\rho_N(x)\right| \leq \left|f(u) - \int_I f(x) d\rho_N(x)\right|$$

$$= \left|\int_I (f(u) - f(x)) d\rho_N(x)\right| \leq \operatorname{var} f,$$

whence

$$|f| = \sup_{u \in I} |f(u)| \leq \left|\int_I f(x) d\rho_N(x)\right| + \operatorname{var} f, \quad f \in BV(I). \qquad (4.5.15)$$

Finally, (4.5.11), (4.5.12), and (4.5.15) imply that

$$\left|U_N^n f - U_N^\infty f\right| \leq \left|\int_I \left(U_N^n f - U_N^\infty f\right)(x) d\rho_N(x)\right| + \operatorname{var}\left(U_N^n f - U_N^\infty f\right)$$

$$\leq \left|U_N^\infty U_N^n f - U_N^\infty f\right| + \operatorname{var} U_N^n f = \operatorname{var} U_N^n f, \qquad (4.5.16)$$

for all $n \in \mathbb{N}$ and $f \in BV(I)$.

It follows from Proposition 4.5.2 that for all $f \in BV(I)$ we have

$$\operatorname{var}\left(U_N f - U_N^\infty f\right) \leq \frac{1}{N} \cdot \operatorname{var}\left(f - U_N^\infty f\right) + K_N \cdot \left|f - U_N^\infty f\right|.$$

But,

$$\operatorname{var}\left(U_N f - U_N^\infty f\right) = \operatorname{var} U_N f, \quad \operatorname{var}\left(f - U_N^\infty f\right) = \operatorname{var} f,$$

and $\left|f - U_N^\infty f\right| \leq \operatorname{var} f$, which is (4.5.16) with $n = 0$. Thus,

$$\operatorname{var} U_N f \leq \frac{1}{N} \cdot \operatorname{var} f + K_N \cdot \operatorname{var} f = \left(\frac{1}{N} + K_N\right) \cdot \operatorname{var} f,$$

which leads to (4.5.13). Next, (4.5.14) follows from (4.5.16) and (4.5.13). □

4.6 The Gauss-Kuzmin-Type Theorem for the Rényi-Type Continued Fraction Expansions

In what follows we give the RSCC associated with the dynamical system $(I, \mathcal{B}_I, R_N)$. The ergodic behavior of this RSCC gives us the asymptotic behavior of R_N^{-n} as $n \to \infty$ that represents the Gauss-Kuzmin-type problem for Rényi-type continued fraction expansions. Proposition 4.3.2 leads us to the RSCC

$$\{(I, \mathcal{B}_I), (\mathbb{N}_N, \mathcal{P}(\mathbb{N}_N)), u_N, P_N\}, \qquad (4.6.1)$$

where $u_N(x, i) := u_{N,i}(x)$ and $P_N(x, i) := P_{N,i}(x)$, $x \in I$, $i \in \mathbb{N}_N$, are given in (4.4.1) and (4.3.9). Let $\mathcal{P}(\mathbb{N}_N)$ denote the power set of $\mathbb{N}_N$.

Proposition 4.6.1 ([42]) *The system* $\{(I, \mathcal{B}_I), (\mathbb{N}_N, \mathcal{P}(\mathbb{N}_N)), u_N, P_N\}$ *is an RSCC with contraction.*

Proof We have

$$\frac{d}{dx} u_{N,i}(x) = \frac{N}{(x+i)^2} \quad \text{and} \quad \frac{d}{dx} P_{N,i}(x) = \frac{i-N}{(x+i-1)^2} - \frac{i+1-N}{(x+i)^2}$$

for any $x \in I$ and $i \in \mathbb{N}_N$. Thus,

$$\sup_{x \in I} \left| \frac{d}{dx} u_{N,i}(x, i) \right| \leq \frac{N}{i^2}, \ i \in \mathbb{N}_N, \quad \sup_{x \in I} \left| \frac{d}{dx} P_{N,i}(x, i) \right| < \infty, \ i \in \mathbb{N}_N.$$

Hence the requirements of Definition 1.4.8 are fulfilled. □

Whatever $a \in I$ the Markov chain $(s_{N,n}^a)_{n \in \mathbb{N}_+}$ associated with RSCC (4.6.1) has the transition operator U_N in (4.5.2) with the transition probability function defined as

$$Q_N(x, B) = \sum_{\{i \in \mathbb{N}_N : u_{N,i}(x) \in B\}} P_{N,i}(x), \quad x \in I, B \in \mathcal{B}_I. \qquad (4.6.2)$$

Proposition 4.6.2 ([42]) *The RSCC (4.6.1) has a regular associated Markov chain.*

Proof By Definition 1.4.2(v), the Markov chain corresponding to RSCC (4.6.1) is regular if its transition operator U_N is regular with respect to $L(I)$. Now, in order to prove the regularity of U_N with respect to $L(I)$, we have, according to Theorem 1.4.9(i), to find an element $x^* \in [0, 1)$ such that $\lim_{n \to \infty} \text{dist}(\sigma_n(x), x^*) = 0$ for all $x \in I$. Here $\sigma_n(w)$ denotes the support of measure $Q_N^n(w, \cdot)$, where Q_N^n is the n-step transition probability function of the Markov chain $(s_{N,n}^a)_{n \in \mathbb{N}_+}$.

Fix $x \in I$. Let us define the sequence $(x_n)_{n \geq 0}$ in I, recursively by

$$x_0 := x, \quad x_{n+1} := 1 - \frac{N}{x_n + N + 1}, \quad n \geq 1. \tag{4.6.3}$$

It is clear that $x_n \in (0, 1)$, $n \in \mathbb{N}$, and letting $n \to \infty$ in (4.6.3), we get

$$0 < x^* = \lim_{n \to \infty} x_n = \frac{\sqrt{N^2 + 4} - N}{2} < 1. \tag{4.6.4}$$

Clearly, $x_{n+1} \in \sigma_1(x_n)$ and using Theorem 1.4.9(ii) and an induction argument lead us to the conclusion that $x_n \in \sigma_n(x)$ for $n \in \mathbb{N}_+$. Since $\text{dist}(\sigma_n(x), x^*) \leq |x_n - x^*| \to 0$ as $n \to \infty$, we obtain that U_N is regular. □

Proposition 4.6.3 ([42]) *The RSCC (4.6.1) is ergodic.*

Proof Since the RSCC (4.6.1) is with contraction and has a regular associated Markov chain, then the statement holds. □

Since the compact Markov chain corresponding to RSCC (4.6.1) is ergodic, Theorem 1.4.6(ii) allows us to obtain the unique stationary probability measure Q_N^∞ on $(I, \mathcal{B}_I)$.

Proposition 4.6.4 ([42]) *The probability Q_N^∞ is the invariant probability measure of the transformation R_N.*

Proof For ρ_N in (4.2.6) and Q_N in (4.6.2), and on account of the uniqueness of Q_N^∞ we have to show that

$$\int_0^1 Q_N(x, B) \, d\rho_N(x) = \rho_N(B), \quad B \in \mathcal{B}_I. \tag{4.6.5}$$

Since the intervals $(u, 1] \subset I$ generate $\mathcal{B}_I$, it is sufficient to show Eq. (4.6.5) just for $B = (u, 1]$, $0 \leq u < 1$. Now, $u_{N,i} \in B$ if and only if $i \geq \lfloor \frac{N}{1-u} - x \rfloor + 1$. Thus,

$$Q_N(x, (u, 1]) = \sum_{\{i \in \mathbb{N}_N : u < u_{N,i}(x) \leq 1\}} P_{N,i}(x) = \sum_{i \geq \lfloor \frac{N}{1-u} - x \rfloor + 1} P_{N,i}(x)$$

$$= \frac{x + N - 1}{x + \lfloor \frac{N}{1-u} - x \rfloor + 1}. \tag{4.6.6}$$

4.6 The Gauss-Kuzmin-Type Theorem for the Rényi-Type Continued...

Then,

$$\int_0^1 Q_N(x,(u,1])d\rho_N(x) = \frac{1}{\log\left(\frac{N}{N-1}\right)} \int_0^1 \frac{dx}{x + \lfloor \frac{N}{1-u} - x \rfloor}$$

$$= \frac{1}{\log\left(\frac{N}{N-1}\right)} \left(\int_0^{\frac{N}{1-u} - \lfloor \frac{N}{1-u} \rfloor} \frac{dx}{x + \lfloor \frac{N}{1-u} \rfloor} \right.$$

$$\left. + \int_{\frac{N}{1-u} - \lfloor \frac{N}{1-u} \rfloor}^1 \frac{dx}{x + \lfloor \frac{N}{1-u} - 1 \rfloor} \right)$$

$$= \frac{1}{\log\left(\frac{N}{N-1}\right)} \log \frac{N}{u + N - 1} = \rho_N((u,1]).$$

Hence the statement holds. □

Note that the Markov operator U_N associated with the RSCC (4.6.1) is also given by

$$U_N f(x) = \int_0^1 f(y) Q_N(x, dy) \quad (4.6.7)$$

with Q_N as in (4.6.2), which implies that

$$U_N^n f(x) = \int_0^1 f(y) Q_N^n(x, dy), \quad x \in I, \ n \in \mathbb{N}_+. \quad (4.6.8)$$

Furthermore by (1.4.12), there exist two positive constants $q_N < 1$ and k_N such that

$$\|U_N^n f - U_N^\infty f\|_L \leq k_N q_N^n \|f\|_L, \quad n \in \mathbb{N}_+, \ f \in L(I), \quad (4.6.9)$$

where

$$U_N^\infty f = \int_0^1 f(y) Q_N^\infty(dy). \quad (4.6.10)$$

Now we are able to show the main result of the section.

Theorem 4.6.5 (A Gauss-Kuzmin-Type Theorem for R_N [42]) *Fix an integer $N \geq 2$ and let $(I, \mathcal{B}_I, R_N)$ be as in Sect. 4.2. For a nonatomic probability measure μ which has* a Riemann-integrable density on $(I, \mathcal{B}_I)$, the following holds:

$$\lim_{n\to\infty} \mu(R_N^n < x) = \frac{1}{\log\left(\frac{N}{N-1}\right)} \log \frac{x+N-1}{N-1}, \quad x \in I. \tag{4.6.11}$$

Proof By (4.5.3), we have

$$\mu\left(R_N^{-n}(A)\right) = \int_A \frac{U_N^n f_{N,0}(x)}{x+N-1} dx \quad \text{for any } n \in \mathbb{N}_+, A \in \mathcal{B}_I, \tag{4.6.12}$$

where $f_{N,0}(x) = (x+N-1)(d\mu/d\lambda)(x)$ for $x \in I$. If $d\mu/d\lambda \in L(I)$, then $f_{N,0} \in L(I)$, and by (4.6.10) and Proposition 4.6.4 we have

$$U_N^\infty f_{N,0} = \int_0^1 f_{N,0}(x) Q_N^\infty(dx) = \int_0^1 f_{N,0}(x) \rho_N(dx) = \frac{1}{\log\left(\frac{N}{N-1}\right)}. \tag{4.6.13}$$

Taking into account (4.6.9), $\|U_N^n f - U_N^\infty f\|_L = \mathcal{O}(q_N^n)\|f\|_L$, as $n \to \infty$ for some positive constant $q_N < 1$, the constant implied in $\mathcal{O}$ being independent of $f \in L(I)$.

Furthermore, consider the Banach space $C(I)$ of all real-valued continuous functions on I with the norm $\|f\| := \sup_{x\in I} |f(x)|$. Since $L(I)$ is a dense subspace of $C(I)$, we have

$$\lim_{n\to\infty} \left\|\left(U_N^n - U_N^\infty\right) f\right\| = 0 \quad \text{for all } f \in C(I). \tag{4.6.14}$$

Therefore, (4.6.14) is valid for a measurable function f which is Q_N^∞-almost surely continuous, that is, for a Riemann-integrable function. Thus, we have

$$\lim_{n\to\infty} \mu\left(R_N^n < x\right) = \lim_{n\to\infty} \int_0^x \frac{U_N^n f_{N,0}(u)}{u+N-1} du = \int_0^x \frac{U_N^\infty f_{N,0}(u)}{u+N-1} du$$
$$= \frac{1}{\log\left(\frac{N}{N-1}\right)} \int_0^x \frac{du}{u+N-1} \doteq \frac{1}{\log\left(\frac{N}{N-1}\right)} \log \frac{x+N-1}{N-1}.$$

□

4.7 Szüsz's Method to Gauss-Kuzmin-Lévy-Type Theorem

In this section we continue our investigation on the asymptotic behavior of the distribution functions of the Rényi-type transformations (see [43]). In order to prove a Gauss-Kuzmin-Lévy-type theorem for the Rényi-type continued fraction expansions, we apply the method of Szüsz [73]. We mention that using this method, we obtain more information on the

4.7 Szüsz's Method to Gauss-Kuzmin-Lévy-Type Theorem

convergence rate involved. The main novelty of this method is the explicit expression in terms of Hurwitz zeta functions of η_N that appears in Theorem 4.7.1. In addition, the estimate we have for η_N shows that $\eta_N \to 0$ as $N \to \infty$.

Let μ be a nonatomic probability measure on $\mathcal{B}_I$ and define

$$F_{N,0}(x) := \mu([0, x]), \quad x \in I, \tag{4.7.1}$$

$$F_{N,n}(x) := \mu(R_N^n \leq x), \quad x \in I, \; n \in \mathbb{N}_+. \tag{4.7.2}$$

Then the following holds.

Theorem 4.7.1 (A Gauss-Kuzmin-Lévy-Type Theorem [43]) *Let R_N and $F_{N,n}$ be as in (4.2.1) and (4.7.2). Then there exists a constant $0 < \eta_N < 1$ such that $F_{N,n}$ can be written as*

$$F_{N,n}(x) = \frac{1}{\log\left(\frac{N}{N-1}\right)} \log\left(\frac{x+N-1}{N-1}\right) + \mathcal{O}(\eta_N^n) \tag{4.7.3}$$

uniformly with respect to $x \in I$.

Remark 4.7.2 ([43]) From (4.7.3), we see that

$$\lim_{n \to \infty} F_{N,n}(x) = \rho_N([0, x]), \tag{4.7.4}$$

where ρ_N is the measure defined in (4.2.6). In fact, Theorem 4.7.1 estimates the error

$$e_{N,n}(\mu; x) = \mu(R_N^n \leq x) - \rho_N([0, x]), \quad x \in I. \tag{4.7.5}$$

To prove Theorem 4.7.1 we need the following results.

Lemma 4.7.3 ([43]) *For functions $\{F_{N,n}\}$ in (4.7.2), the following Gauss-Kuzmin-type equation holds:*

$$F_{N,n+1}(x) = \sum_{i \geq N} \left\{ F_{N,n}\left(1 - \frac{N}{x+i}\right) - F_{N,n}\left(1 - \frac{N}{i}\right) \right\} \tag{4.7.6}$$

for $x \in I$ and $n \in \mathbb{N}$.

Proof From (4.2.1) and (4.2.2), we see that

$$R_N^n(x) = 1 - \frac{N}{r_{n+1} + R_N^{n+1}}, \quad n \in \mathbb{N}_+. \tag{4.7.7}$$

Now,

$$F_{N,n+1}(x) = \mu\left(R_N^{n+1} \leq x\right) = \sum_{i \geq N} \mu\left(1 - \frac{N}{i} \leq R_N^n \leq 1 - \frac{N}{i+x}\right)$$

$$= \sum_{i \geq N} \left\{F_{N,n}\left(1 - \frac{N}{x+i}\right) - F_{N,n}\left(1 - \frac{N}{i}\right)\right\}.$$

□

Remark 4.7.4 ([43]) Assume that for some $p \in \mathbb{N}$, the derivative $F'_{N,p}$ exists everywhere in I and is bounded. Then it is easy to see by induction that $F'_{N,p+n}$ exists and is bounded for all $n \in \mathbb{N}_+$. This allows us to differentiate (4.7.6) term by term, obtaining

$$F'_{N,n+1}(x) = \sum_{i \geq N} \frac{N}{(x+i)^2} F'_{N,n}\left(1 - \frac{N}{x+i}\right). \tag{4.7.8}$$

We introduce functions $\{f_{N,n}\}$ as follows:

$$f_{N,n}(x) := (x + N - 1) F'_{N,n}(x), \quad x \in I, \ n \in \mathbb{N}. \tag{4.7.9}$$

Then (4.7.8) is

$$f_{N,n+1}(x) = \sum_{i \geq N} P_{N,i}(x) f_{N,n}\left(u_{N,i}(x)\right), \tag{4.7.10}$$

where $P_{N,i}(x)$ and $u_{N,i}(x)$ are given in (4.3.9) and (4.4.1), respectively.

Lemma 4.7.5 ([43]) *For $\{f_{N,n}\}$ in (4.7.9), define $M_{N,n} := \max_{x \in I} |f'_{N,n}(x)|$. Then*

$$M_{N,n+1} \leq \eta_N \cdot M_{N,n}, \tag{4.7.11}$$

where

$$\eta_N := \sum_{i \geq N} \left(\frac{1}{i^3} + \frac{N}{i^2(i+1)}\right). \tag{4.7.12}$$

Proof Since

$$P_{N,i}(x) = \frac{i+1-N}{x+i} - \frac{i-N}{x+i-1},$$

4.7 Szüsz's Method to Gauss-Kuzmin-Lévy-Type Theorem

we have

$$f'_{N,n+1}(x) = \sum_{i \geq N} \left\{ P'_{N,i}(x) f_{N,n}\left(u_{N,i}(x)\right) + P_{N,i}(x) f'_{N,n}\left(u_{N,i}(x)\right) u'_{N,i}(x) \right\}$$

$$= \sum_{i \geq N} \left\{ \left(\frac{i-N}{(x+i-1)^2} - \frac{i+1-N}{(x+i)^2} \right) f_{N,n}\left(u_{N,i}(x)\right) \right.$$

$$\left. + P_{N,i}(x) f'_{N,n}\left(u_{N,i}(x)\right) \frac{N}{(x+i)^2} \right\}$$

$$= \sum_{i \geq N} \left\{ \frac{i+1-N}{(x+i)^2} \left[f_{N,n}\left(u_{N,i+1}(x)\right) - f_{N,n}\left(u_{N,i}(x)\right) \right] \right.$$

$$\left. + P_{N,i}(x) f'_{N,n}\left(u_{N,i}(x)\right) \frac{N}{(x+i)^2} \right\}$$

$$= \sum_{i \geq N} \left\{ \frac{i+1-N}{(x+i)^3(x+i+1)} f'_{N,n}(\theta_i) + f'_{N,n}\left(u_{N,i}(x)\right) \frac{N P_{N,i}(x)}{(x+i)^2}, \right\} \quad (4.7.13)$$

where $u_{N,i+1}(x) < \theta_i < u_{N,i}(x)$. Now (4.7.13) implies

$$M_{N,n+1} \leq M_{N,n} \cdot \max_{x \in I} \left(\sum_{i \geq N} \frac{i+1-N}{(x+i)^3(x+i+1)} + N \sum_{i \geq N} \frac{x+N-1}{(x+i)^3(x+i-1)} \right). \quad (4.7.14)$$

We now must calculate the maximum value of the sums in this expression. Using that $x \in I$ and $i \geq N$, we get

$$\frac{i+1-N}{(x+i)^3(x+i+1)} \leq \frac{i+1-N}{i^3(i+1)} \quad \text{and} \quad \frac{x+N-1}{(x+i)^3(x+i-1)} \leq \frac{1}{(x+i)^3} \leq \frac{1}{i^3}.$$

Thus,

$$M_{N,n+1} \leq M_{N,n} \cdot \sum_{i \geq N} \left(\frac{1}{i^3} + \frac{N}{i^2(i+1)} \right) \quad (4.7.15)$$

and the proof is complete. □

Proof of Theorem 4.7.1. Let $\{F_{N,n}\}$ and $\{f_{N,n}\}$ be as in (4.7.2) and (4.7.9), respectively. Then

$$F'_{N,n}(x) = \frac{1}{x+N-1} f_{N,n}(x), \quad x \in I, n \in \mathbb{N}. \quad (4.7.16)$$

If we can show the existence of a constant $0 < \eta_N < 1$ such that

$$f_{N,n}(x) = \frac{1}{\log\left(\frac{N}{N-1}\right)} + \mathcal{O}(\eta_N^n), \qquad (4.7.17)$$

then integrating (4.7.16) we will establish (4.7.3). To demonstrate that $f_{N,n}(x)$ has the desired form, it suffices to establish that $f'_{N,n}(x) = \mathcal{O}\left(\eta_N^n\right)$, as the $\frac{1}{\log\left(\frac{N}{N-1}\right)}$ constant in (4.7.17) will follow from the normalization requirement that $F_{N,n}(0) = 0$ and $F_{N,n}(1) = 1$. □

It remains to prove the following lemma.

Lemma 4.7.6 ([43]) *For every integer $N \geq 2$ there exists a constant $0 < \eta_N < 1$ such that*

$$f'_{N,n}(x) = \mathcal{O}(\eta_N^n), \quad x \in I, \; n \in \mathbb{N}. \qquad (4.7.18)$$

Moreover, for any positive integer $N \geq 2$ the following estimate holds

$$\frac{1}{N^3} + \frac{1}{2N(N+1)} + \frac{1}{2N} < \eta_N < \frac{1}{2N(N-1)} + \frac{1}{N} - \frac{1}{2N+1}. \qquad (4.7.19)$$

Proof Let η_N be as in Lemma 4.7.5. Using this lemma, to show (4.7.18) it is enough to prove that $\eta_N < 1$. First, we will write η_N in terms of Hurwitz zeta functions. Thus,

$$\eta_N = \sum_{i \geq N} \left(\frac{1}{i^3} + \frac{N}{i^2(i+1)}\right) = \sum_{i \geq N} \left(\frac{1}{i^3} + \frac{N}{i^2}\right) - 1 = \zeta(3, N) + N\zeta(2, N) - 1.$$

For $i \geq N$ and $a_N := \frac{1}{2}\left(\sqrt{4N^2+1} - (2N+1)\right) > -\frac{1}{2}$, we have

$$a_N^2 + (2N+1)a_N + N = 0 \text{ and } a_N^2 + (2N+1)a_N + N + (2a_N+1)(i-N) \geq 0,$$

i.e., $i^2 \leq (i+a_N)(i+1+a_N)$. Hence,

$$\zeta(2, N) \geq \sum_{i \geq N} \left(\frac{1}{i+a_N} - \frac{1}{i+1+a_N}\right) = \frac{1}{N+a_N} = \frac{2}{\sqrt{4N^2+1}-1}.$$

Also we have

$$\zeta(2, N) < \frac{1}{N^2} + \sum_{i \geq N+1} \frac{1}{(i-1/2)(i+1/2)} = \frac{1}{N^2} + \frac{2}{2N+1}.$$

4.7 Szüsz's Method to Gauss-Kuzmin-Lévy-Type Theorem

For $i \geq N$ and $b_N := N\left(\sqrt{N^2+1} - N\right) < \frac{1}{2}$, we have

$$b_N^2 + 2N^2 b_N - N^2 = 0 \text{ and } b_N^2 + 2N^2 b_N - N^2 + (2b_N - 1)(i^2 - N^2) \leq 0,$$

i.e., $i^4 \geq (i^2 - i + b_N)(i^2 + i + b_N)$. Hence

$$\zeta(3, N) < \frac{1}{2} \sum_{i \geq N} \left(\frac{1}{(i-1)i + b_N} - \frac{1}{i(i+1) + b_N}\right) = \frac{1}{2(N^2 - N + b_N)}$$

$$= \frac{1}{2N(\sqrt{N^2+1} - 1)}.$$

Also, we have

$$\zeta(3, N) > \frac{1}{N^3} + \frac{1}{2} \sum_{i \geq N+1} \left(\frac{1}{i^2 - i + 1/2} + \frac{1}{i^2 + i + 1/2}\right) = \frac{1}{N^3} + \frac{1}{2(N^2 + N + 1/2)}.$$

Therefore,

$$\eta_N < \frac{1}{2N(\sqrt{N^2+1} - 1)} + N\left(\frac{1}{N^2} + \frac{2}{2N+1}\right) - 1 < \frac{1}{2N(N-1)} + \frac{1}{N} - \frac{1}{2N+1},$$

and

$$\eta_N > \frac{1}{N^3} + \frac{1}{2(N^2 + N + 1/2)} + \frac{2N}{\sqrt{4N^2 + 1} - 1} - 1 > \frac{1}{N^3} + \frac{1}{2(N^2 + N + 1/2)}$$

$$+ \frac{2N+1}{2N} - 1 = \frac{1}{N^3} + \frac{1}{2(N^2 + N + 1/2)} + \frac{1}{2N} > \frac{1}{N^3} + \frac{1}{2N(N+1)} + \frac{1}{2N}.$$

For example, we have (Table 4.1):

Table 4.1 Lower and upper bounds of error for some $N \geq 1$

N	Lower bound of η_N	Upper bound of η_N
2	0.4583333333333333	0.55
10	0.055545454545454544	0.057936507936507944
100	0.00505050495049505	0.0050753806723955975
500	0.001002004007984032	0.001003003009015033
1000	0.0005005005004995005	0.0005007503755629693
5000	0.00010002000400079984	0.00010003000300090015
10000	0.00005000500050004999	0.000050007500375056254

□

4.8 Wirsing-Type Approach to the Perron-Frobenius Operator

The Gauss-Kuzmin-Lévy problem for the transformation R_N can be approached in terms of the Perron-Frobenius operator under the invariant measure ρ_N induced by the limit distribution function. By restricting the domain of this operator to the Banach space of functions which have a continuous derivative on I, we obtained in [67] upper and lower bounds of the error which provide a refined estimate of the convergence rate. For example, in the case $N = 100$, the upper and lower bounds of the convergence rate are respectively $\mathcal{O}\left(w_{100}^n\right)$ and $\mathcal{O}\left(v_{100}^n\right)$ as $n \to \infty$, with $v_{100} > 0.00503350150708559$ and $w_{100} < 0.00503358526129032$.

4.8.1 Near-Optimal Solution to the Gauss-Kuzmin-Lévy Problem

In this section we develop a Wirsing-type approach [76] to obtain a solution to the Gauss-Kuzmin-Lévy problem in Theorem 4.8.4.

Let μ be a probability measure on $\mathcal{B}_I$ such that $\mu \ll \lambda$. For any $n \in \mathbb{N}$ put $F_{N,n}(x) = \mu\left(R_N^n < x\right)$, $x \in I$, where R_N^0 is the identity map. As $\left(R_N^n < x\right) = R_N^{-n}((0,x))$, by Proposition 4.5.1 (ii), we have

$$F_{N,n}(x) = \int_0^x \frac{U_N^n f_{N,0}(u)}{u + N - 1} du, \quad n \in \mathbb{N}, \tag{4.8.1}$$

where $f_{N,0}(x) := (x + N - 1)\left(F_{N,0}\right)'(x)$, $x \in I$, where $\left(F_{N,0}\right)' = d\mu/d\lambda$.

We will assume that $\left(F_{N,0}\right)' \in C^1(I)$, the collection of all functions $f : I \to \mathbb{C}$ which have a continuous derivative. So, we study the behavior of U_N^n as $n \to \infty$, assuming that the domain of U_N is $C^1(I)$.

Let $f \in C^1(I)$. Then the series (4.5.2) can be differentiated term by term, since the series of derivatives is uniformly convergent. Next, since

$$P_{N,i}(x) = \frac{i + 1 - N}{x + i} - \frac{i - N}{x + i - 1},$$

we get

$$(U_N f)'(x) = \sum_{i \geq N} \left\{ (P_{N,i})'(x) f\left(u_{N,i}(x)\right) + P_{N,i}(x) f'\left(u_{N,i}(x)\right) \left(u_{N,i}\right)'(x) \right\}$$

$$= \sum_{i \geq N} \left\{ \left(\frac{i - N}{(x + i - 1)^2} - \frac{i + 1 - N}{(x + i)^2} \right) f\left(u_{N,i}(x)\right) + P_{N,i}(x) f'\left(u_{N,i}(x)\right) \frac{N}{(x + i)^2} \right\}$$

$$= \sum_{i \geq N} \left\{ \frac{i + 1 - N}{(x + i)^2} \left[f\left(u_{N,i+1}(x)\right) - f\left(u_{N,i}(x)\right) \right] + P_{N,i}(x) f'\left(u_{N,i}(x)\right) \frac{N}{(x + i)^2} \right\}$$

$$\tag{4.8.2}$$

4.8 Wirsing-Type Approach to the Perron-Frobenius Operator

for any $x \in I$. Thus, we can write

$$(U_N f)' = -U_N^* f', \quad f \in C^1(I), \tag{4.8.3}$$

where $U_N^* : C(I) \to C(I)$ is defined by

$$U_N^* g(x) = -\sum_{i \geq N} \left\{ \frac{i+1-N}{(x+i)^2} \int_{u_{N,i}(x)}^{u_{N,i+1}(x)} g(u) du + \frac{N(x+N-1)}{(x+i-1)(x+i)^3} g\left(u_{N,i}(x)\right) \right\} \tag{4.8.4}$$

with $g \in C(I)$ and $x \in I$. Clearly, $(U_N^n f)' = (-1)^n (U_N^*)^n f'$, $n \in \mathbb{N}_+$, $f \in C^1(I)$. We are going to show that $(U_N^*)^n$ takes certain functions into functions with very small values when $n \in \mathbb{N}_+$ is large.

Proposition 4.8.1 ([67]) *For a fixed integer $N \geq 2$ there are positive constants $v_N < w_N < 1$ and a real-valued nonpositive function $\varphi_N \in C(I)$ such that*

$$v_N(-\varphi_N) \leq U_N^* \varphi_N \leq w_N(-\varphi_N). \tag{4.8.5}$$

Proof For $\mathbb{R}_+ := \{x \in \mathbb{R} : x \geq 0\}$, let $h_N : \mathbb{R}_+ \to \mathbb{R}$ be a continuous bounded function such that $\lim_{x \to \infty} h_N(x) < \infty$. We look for a function $g_N : (0,1] \to \mathbb{R}$ such that $U_N g_N = h_N$, assuming that the equation

$$U_N g_N(x) = \sum_{i \geq N} P_{N,i}(x) g_N\left(u_{N,i}(x)\right) = h_N(x) \tag{4.8.6}$$

holds for $x \in \mathbb{R}_+$. Reducing the terms of the series involved in (4.8.6) yields

$$\frac{h_N(x)}{x+N-1} - \frac{h_N(x+1)}{x+N} = \frac{1}{(x+N-1)(x+N)} g_N\left(\frac{x}{x+N}\right), \quad x \in \mathbb{R}_+. \tag{4.8.7}$$

Hence

$$g_N(u) = \frac{N}{1-u} h_N\left(\frac{Nu}{1-u}\right) - \left(\frac{N}{1-u} - 1\right) h_N\left(\frac{Nu}{1-u} + 1\right), \quad u \in (0,1], \tag{4.8.8}$$

and we indeed have $U_N g_N = h_N$ since

$$U_N g_N(x) = \sum_{i \geq N} \frac{x+N-1}{(x+i-1)(x+i)} g_N\left(1 - \frac{N}{x+i}\right)$$

$$= (x+N-1) \sum_{i \geq N} \left(\frac{1}{x+i-1} - \frac{1}{x+i}\right) g_N\left(1 - \frac{N}{x+i}\right)$$

$$= (x+N-1) \sum_{i \geq N} \left(\frac{h_N(x+i-N)}{x+i-1} - \frac{h_N(x+i+1-N)}{x+i} \right)$$

$$= (x+N-1) \left(\frac{h_N(x)}{x+N-1} - \lim_{i \to \infty} \frac{h_N(x+i+1-N)}{x+i} \right) = h_N(x), \quad x \in \mathbb{R}_+.$$

In particular, for any fixed $t_N \in I$ we consider the function $h_{N,t_N} : \mathbb{R}_+ \to \mathbb{R}$ defined as

$$h_{N,t_N}(x) := \frac{1}{e_N x + t_N + 1}, \quad x \in \mathbb{R}_+, \tag{4.8.9}$$

where the coefficient e_N will be specified later. By the above, the function $g_{N,t_N} : (0, 1] \to \mathbb{R}$ defined as

$$g_{N,t_N}(x) = \frac{N}{1-x} h_{N,t_N} \left(\frac{Nx}{1-x} \right) - \left(\frac{N}{1-x} - 1 \right) h_{N,t_N} \left(\frac{Nx}{1-x} + 1 \right)$$

$$= \frac{N}{e_N N x + (t_N + 1)(1-x)} - \frac{N-1+x}{e_N(Nx+1-x) + (t_N+1)(1-x)} \tag{4.8.10}$$

for any $x \in (0, 1]$ satisfies $U g_{N,t_N}(x) = h_{N,t_N}(x)$, $x \in I$. Setting

$$\varphi_{N,t_N}(x) := (g_{N,t_N})'(x)$$

$$= \frac{-N(e_N N - t_N - 1)}{\{e_N N x + (t_N+1)(1-x)\}^2} - \frac{N\{2e_N - (e_N N - t_N - 1)\}}{\{e_N(Nx+1-x) + (t_N+1)(1-x)\}^2}, \tag{4.8.11}$$

we have

$$U_N^* \varphi_{N,t_N}(x) = -(U g_{N,t_N})'(x) = -(h_{N,t_N})'(x) = \frac{e_N}{(e_N x + t_N + 1)^2}, \quad x \in I. \tag{4.8.12}$$

Since g_{N,t_N} is a decreasing function, it follows that $\varphi_{N,t_N}(x) < 0$, $x \in I$. Also, U_N^* is a linear operator that takes nonpositive functions into positive functions. Therefore, $U_N^* \varphi_{N,t_N}(x) > 0, x \in I$.

We choose t_N by asking that $(\varphi_{N,t_N}/U_N^*\varphi_{N,t_N})(0) = (\varphi_{N,t_N}/U_N^*\varphi_{N,t_N})(1)$. Since

$$(\varphi_{N,t_N}/U_N^*\varphi_{N,t_N})(0) = \frac{(t_N+1)^2}{e_N} \left\{ \frac{-N(e_N N - t_N - 1)}{(t_N+1)^2} - \frac{N\{2e_N-(e_N N-t_N-1)\}}{(e_N+t_N+1)^2} \right\} \tag{4.8.13}$$

and

$$(\varphi_{N,t_N}/U_N^*\varphi_{N,t_N})(1) = \frac{-2(e_N+t_N+1)^2}{e_N^2 N}, \tag{4.8.14}$$

4.8 Wirsing-Type Approach to the Perron-Frobenius Operator

this amounts to the equation

$$H_N(t_N) = 2(t_N + e_N + 1)^4 + e_N^3 N^2(1 - 2N)(t_N + 1) - e_N^4 N^3 = 0. \tag{4.8.15}$$

We choose the coefficient e_N such that the equation $H_N(x) = 0$, $x \in I$, yields a unique solution $t_N \in I$. Asking that $H_N(0) < 0$, $H_N(1) > 0$, and $\dfrac{dH_N}{dt_N} > 0$, we may determine e_N (see Sect. 4.8.2). For this unique acceptable solution $t_N \in I$ the function $\varphi_{N,t_N}/U_N^*\varphi_{N,t_N}$ attains its minimum equal to $-2(e_N + t_N + 1)^2/(e_N^2 N)$ at $x = 0$ and $x = 1$ and has a maximum $m(t_N) = \varphi_{N,t_N}/U_N^*\varphi_{N,t_N}(x_{\max}) < 0$. It follows that

$$\frac{-2(e_N + t_N + 1)^2}{e_N^2 N} \leq \frac{\varphi_{N,t_N}}{U_N^*\varphi_{N,t_N}} \leq m(t_N).$$

Since $\dfrac{\varphi_{N,t_N}}{U_N^*\varphi_{N,t_N}} < 0$, we get

$$-m(t_N) \leq \frac{-\varphi_{N,t_N}}{U_N^*\varphi_{N,t_N}} \leq \frac{2(e_N + t_N + 1)^2}{e_N^2 N}.$$

Therefore,

$$\frac{e_N^2 N}{2(e_N + t_N + 1)^2}(-\varphi_{N,t_N}) \leq U_N^*\varphi_{N,t_N} \leq -\frac{1}{m(t_N)}(-\varphi_{N,t_N}).$$

It follows that for $\varphi_N = \varphi_{N,t_N}$ we have

$$v_N(-\varphi_N) \leq U_N^*\varphi_N \leq w_N(-\varphi_N), \text{ where } v_N = \frac{e_N^2 N}{2(e_N + t_N + 1)^2} \text{ and } w_N = -\frac{1}{m(t_N)}.$$

$\square$

Remark 4.8.2 ([67]) By (4.8.5) we successively get

$$v_N^2(-\varphi_N) \leq -(U_N^*)^2\varphi_N \leq w_N^2(-\varphi_N)$$
$$v_N^3(-\varphi_N) \leq (U_N^*)^3\varphi_N \leq w_N^3(-\varphi_N)$$
$$\vdots$$
$$v_N^n(-\varphi_N) \leq (-1)^{n+1}(U_N^*)^n\varphi_N \leq w_N^n(-\varphi_N), \quad n \in \mathbb{N}_+.$$

Corollary 4.8.3 ([67]) *Let $f_{N,0} \in C^1(I)$ such that $(f_{N,0})' > 0$. Put*

$$\alpha_N = \min_{x \in [0,1]} \frac{-\varphi_N(x)}{(f_{N,0})'(x)} \text{ and } \beta_N = \max_{x \in [0,1]} \frac{-\varphi_N(x)}{(f_{N,0})'(x)}.$$

Then

$$\frac{\alpha_N}{\beta_N} v_N^n (f_{N,0})'(x) \leq (-1)^n (U_N^*)^n (f_{N,0})'(x) \leq \frac{\beta_N}{\alpha_N} w_N^n (f_{N,0})'(x), \quad n \in \mathbb{N}_+, \ x \in I. \tag{4.8.16}$$

Proof Noting that $\alpha_N (f_{N,0})'(x) \leq (-\varphi_N)(x) \leq \beta_N (f_{N,0})'(x)$ and using Remark 4.8.2 we can write

$$\frac{\alpha_N}{\beta_N} v_N^n (f_{N,0})'(x) \leq \frac{1}{\beta_N} v_N^n (-\varphi_N)(x) \leq \frac{1}{\beta_N} (-1)^{n+1} (U_N^*)^n \varphi_N(x)$$

$$\leq (-1)^n (U_N^*)^n (f_{N,0})'(x)$$

$$\leq \frac{(-1)^{n+1}}{\alpha_N} (U_N^*)^n \varphi_N(x) \leq \frac{w_N^n}{\alpha_N} (-\varphi_N)(x) \leq \frac{\beta_N}{\alpha_N} w_N^n (f_{N,0})'(x), \quad n \in \mathbb{N}_+,$$

which shows that (4.8.16) holds. □

Theorem 4.8.4 ([67]) *Let $f_{N,0} \in C^1(I)$ such that $(f_{N,0})' > 0$ and let μ be a probability measure on $\mathcal{B}_I$ such that $\mu \ll \lambda$. For any $n \in \mathbb{N}_+$ and $x \in I$ we have*

$$\left(\log \left(\frac{N}{N-1} \right) \right)^2 \cdot \frac{N}{2} \cdot \frac{\alpha_N}{\beta_N} \min_{x \in [0,1]} (f_{N,0})'(x) \cdot v_N^n F_N(x)(1 - F_N(x))$$

$$\leq |\mu(R_N^n < x) - F_N(x)|$$

$$\leq \left(\log \left(\frac{N}{N-1} \right) \right)^2 \cdot \frac{N}{2} \cdot \frac{\beta_N}{\alpha_N} \max_{x \in [0,1]} (f_{N,0})'(x) \cdot w_N^n F_N(x)(1 - F_N(x)), \tag{4.8.17}$$

where α_N, β_N, v_N, and w_N are defined in Proposition 4.8.1 and Corollary 4.8.3, and

$$F_N(x) = \frac{1}{\log \left(\frac{N}{N-1} \right)} \log \left(\frac{x+N-1}{N-1} \right). \tag{4.8.18}$$

Proof For any $n \in \mathbb{N}$ and $x \in I$ set $d_n(F_N(x)) = \mu(R_N^n < x) - F_N(x)$, with F_N as in (4.8.18). Then by (4.8.1) we have

$$d_n(F_N(x)) = \int_0^x \frac{U_N^n f_{N,0}(u)}{u + N - 1} du - F_N(x).$$

4.8 Wirsing-Type Approach to the Perron-Frobenius Operator

Differentiating twice with respect to x yields

$$d'_n(F_N(x)) \frac{1}{\log\left(\frac{N}{N-1}\right)} \frac{1}{x+N-1} = \frac{U_N^n f_{N,0}(x)}{x+N-1} - \frac{1}{\log\left(\frac{N}{N-1}\right)} \frac{1}{x+N-1},$$

$$(U_N^n f_{N,0})'(x) = d''_n(F_N(x)) \left(\frac{1}{\log\left(\frac{N}{N-1}\right)}\right)^2 \frac{1}{x+N-1}, \quad n \in \mathbb{N}, x \in I. \quad (4.8.19)$$

Hence by (4.8.3) we have

$$d''_n(F_N(x)) = \left(\log\left(\frac{N}{N-1}\right)\right)^2 (x+N-1) \left(U_N^n f_{N,0}\right)'(x)$$

$$= (-1)^n \left(\log\left(\frac{N}{N-1}\right)\right)^2 (x+N-1)(U_N^*)^n (f_{N,0})'(x),$$

for any $n \in \mathbb{N}$, $x \in I$. Since $d_n(0) = d_n(1) = 0$, a well-known interpolation formula yields

$$d_n(x) = -\frac{x(1-x)}{2} d''_n(\xi), \quad n \in \mathbb{N}, x \in I, \quad (4.8.20)$$

for a suitable $\xi = \xi(n, x) \in I$. Therefore

$$\mu(R_N^n < x) - F_N(x) = -\frac{F_N(x)(1 - F_N(x))}{2} d''_n(F_N(\xi_N))$$

$$= (-1)^{n+1} \left(\log\left(\frac{N}{N-1}\right)\right)^2 (\xi_N + N - 1)(U_N^*)^n (f_{N,0})'(\xi_N) \frac{F_N(x)(1 - F_N(x))}{2}$$

$$\leq (-1)^{n+1} \left(\log\left(\frac{N}{N-1}\right)\right)^2 \frac{N}{2} (U_N^*)^n (f_{N,0})'(\xi_N) F_N(x)(1 - F_N(x))$$

for any $n \in \mathbb{N}$ and $x \in [0, 1]$, and another suitable $\xi_N = \xi_N(n, x) \in I$. The result stated follows now from Corollary 4.8.3. □

4.8.2 Final Remarks

To conclude, we use the values obtained in Table 4.3 (see [67]).

Let us consider the case $N = 100$. The equation $H_{100}(x) = 0$, with $e_{100} = 0.0152027$, has as unique acceptable solution $t_{100} = 0.4999998$. For this value of t_N the function $\varphi_{t_{100}}/U_N^* \varphi_{t_{100}}$ attains its minimum equal to -198.668858764086 at $x = 0$

Table 4.2 Lower and upper bounds of the convergence rate for some $N \geq 2$

$N = 3$	$v_3 > 0.20967015556054$	$w_3 < 0.216093436628214$
$N = 5$	$v_5 > 0.114674266412028$	$w_5 < 0.115692692356046$
$N = 100$	$v_{100} > 0.00503350150708559$	$w_{100} < 0.00503358526129032$

Table 4.3 The values of e_N and t_N for some $N \geq 2$

N	e_N	t_N
2	2.1780250	0.4999997
3	0.8956735	0.4999967
4	0.5616365	0.5000001
5	0.4088150	0.5000000
10	0.1730660	0.5000007
15	0.109754	0.4999972
20	0.080357	0.5000078
25	0.063380	0.4999999
30	0.052326	0.5000149
40	0.038793	0.4999972
50	0.030822	0.5000046
100	0.0152027	0.4999998
1000	0.00150201	0.5000073
10000	0.00015002	0.4999999

and $x = 1$, and has a maximum $m(t_{100}) = (\varphi_{t_{100}}/U_N^* \varphi_{t_{100}})(0.49804751660470764) = -198.66555309796482$. It follows that upper and lower bounds of the convergence rate are respectively $O(w_{100}^n)$ and $O(v_{100}^n)$ as $n \to \infty$, with $v_{100} > 0.00503350150708559$ and $w_{100} < 0.00503358526129032$ (Table 4.2).

Imposing the necessary conditions and using MATHEMATICA, we obtain (Table 4.3):

4.9 A Two-Dimensional Gauss-Kuzmin Theorem

In this section we study the joint distribution function of R_N^n and $s_{N,n}^a$, $n \in \mathbb{N}_+$, under $\rho_{N,a}$, $a \in I$. We derive the asymptotic distribution function

$$\lim_{n \to \infty} \rho_{N,a}\left(R_N^n \in [0, x], s_{N,n}^a \in [0, y]\right) = \frac{1}{\log\left(\frac{N}{N-1}\right)}$$

$$\times \log \frac{(x + N - 1)(y + N - 1)}{(N - 1)(N - (1 - x)(1 - y))}, x, y \in I,$$

4.9 A Two-Dimensional Gauss-Kuzmin Theorem

and we deliver an estimate of the n-th error term

$$e_{N,n}(\rho_{n,a}; x, y) = \rho_{N,a}\left(R_N^n \in [0, x], s_{N,n}^a \in [0, y]\right) - \frac{1}{\log\left(\frac{N}{N-1}\right)}$$

$$\times \log \frac{(x+N-1)(y+N-1)}{(N-1)(N-(1-x)(1-y))}$$

for any $a \in I$, $x, y \in I$, and $n \in \mathbb{N}_+$.

The main result of this section is Theorem 4.9.5. We shall derive lower and upper bounds (not depending on $a \in I$) of the supremum

$$\sup_{x,y \in I} |e_{N,n}(\rho_{n,a}; x, y)|, \quad a \in I, \tag{4.9.1}$$

which provide an estimate of the convergence rate involved. First, we obtain a lower bound for the error, which suggests the convergence rate of $\rho_{N,a}\left(s_{N,n}^a \in [0, y]\right)$ to $\rho_N([0, y])$ as $n \to \infty$ for all $a \in I$.

Proposition 4.9.1 ([68]) *For any $a \in I$ and $n \in \mathbb{N}_+$ we have*

$$\frac{1}{2} P_{N,N(n)}(1) \leq \sup_{y \in I} \left|\rho_{N,a}\left(s_{N,n}^a \in [0, y]\right) - \rho_N([0, y])\right|$$

with $P_{N,N(n)}(a) = \sup_{s \in I} \rho_{N,a}\left(s_{N,n}^a = s\right)$, where we write $N(n)$ for $(i_1, \ldots, i_n)$ with $i_1 = \ldots = i_n = N$, $n \in \mathbb{N}_+$.

Proof First, the continuity of the function $y \mapsto \rho_N([0, y])$, $y \in I$, and the equations

$$\lim_{h \searrow 0} \rho_{N,a}\left(s_{N,n}^a \leq y - h\right) = \rho_{N,a}\left(s_{N,n}^a < y\right) \text{ and } \lim_{h \searrow 0} \rho_{N,a}\left(s_{N,n}^a < y + h\right)$$

$$= \rho_{N,a}\left(s_{N,n}^a \leq y\right)$$

imply that

$$\sup_{y \in I} \left|\rho_{N,a}\left(s_{N,n}^a \leq y\right) - \rho_N([0, y])\right| = \sup_{y \in I} \left|\rho_{N,a}\left(s_{N,n}^a < y\right) - \rho_N([0, y])\right|$$

for all $a \in I$ and $n \in \mathbb{N}$. Second, whatever $s \in I$ we have

$$\rho_{N,a}(s_{N,n}^a = s) = \rho_{N,a}\left(s_{N,n}^a \leq s\right) - \rho_N([0, s]) - \left(\rho_{N,a}\left(s_{N,n}^a < s\right) - \rho_N([0, s])\right)$$

$$\leq \sup_{y \in I} \left|\rho_{N,a}\left(s_{N,n}^a \leq y\right) - \rho_N([0, y])\right|$$

$$+ \sup_{y \in I} \left| \rho_{N,a} \left(s_{N,n}^a < y \right) - \rho_N \left([0, y] \right) \right|$$

$$= 2 \sup_{y \in I} \left| \rho_{N,a} \left(s_{N,n}^a \leq y \right) - \rho_N \left([0, y] \right) \right|.$$

Hence for all $a \in I$ and $n \in \mathbb{N}$, we have

$$\sup_{y \in I} \left| \rho_{N,a} \left(s_{N,n}^a \in [0, y] \right) - \rho_N \left([0, y] \right) \right| = \sup_{y \in I} \left| \rho_{N,a} \left(s_{N,n}^a \leq y \right) - \rho_N \left([0, y] \right) \right|$$

$$\geq \frac{1}{2} \sup_{s \in I} \rho_{N,a} \left(s_{N,n}^a = s \right).$$

By induction with respect to $n \in \mathbb{N}_+$ we get

$$U_N^n f(x) = \sum_{i_1, \ldots, i_n \geq N} P_{N, i_1 \ldots i_n}(x) f(u_{N, i_n \ldots i_1}(x)), \quad x \in I, \tag{4.9.2}$$

where

$$u_{N, i_n \ldots i_1} := u_{N, i_n} \circ \ldots \circ u_{N, i_1} \tag{4.9.3}$$

$$P_{N, i_1 \ldots i_n}(x) := P_{N, i_1}(x) P_{N, i_2}(u_{N, i_1}(x)) \ldots P_{N, i_n}(u_{N, i_{n-1} \ldots i_1}(x)), \quad n \geq 2, \tag{4.9.4}$$

and the functions $u_{N,i}$ and $P_{N,i}$ are defined in (4.4.1) and (4.3.9), respectively, for all $i \in \mathbb{N}_N$. Next, using (4.4.14), we have

$$U_N^n f(a) = E_{\rho_{N,a}} \left(f \left(s_{N,n}^a \right) \right), \quad n \in \mathbb{N}, f \in B(I), a \in I.$$

As $s_{N,n}^a = u_{N, r_n, \ldots, r_1}(a), a \in I, n \in \mathbb{N}_+$, we have

$$U_N^n f(a) = \sum_{i^{(n)} \in (\mathbb{N}_N)^n} \rho_{N,a} \left((r_1, r_2, \ldots, r_n) = i^{(n)} \right) f \left(u_{N, i_n \ldots i_1}(a) \right) \tag{4.9.5}$$

for any $n \in \mathbb{N}_+, f \in B(I), a \in I$, and $i^{(n)} = (i_1, \ldots, i_n) \in (\mathbb{N}_N)^n$. Hence, by (4.3.1), (4.9.2), and (4.9.5) we get

$$P_{N, i_1 \ldots i_n}(a) = \rho_{N,a} \left(I_R \left(i^{(n)} \right) \right) = \rho_{N,a} \left(s_{N,n}^a = [i_n, \ldots, i_2, i_1 + a - 1]_R \right), \quad n \geq 2,$$

$$P_{N, i_1}(a) = \rho_{N,a} \left(I_R(i_1) \right) = \rho_{N,a} \left(s_{N,1}^a = 1 - \frac{N}{i_1 + a} \right),$$

4.9 A Two-Dimensional Gauss-Kuzmin Theorem

for all $a \in I$ and $i_1, \ldots, i_n \in \mathbb{N}_N$. Since as easily seen,

$$\max_{i^{(n)} \in (\mathbb{N}_N)^n} \rho_{N,a}\left(I_R\left(i^{(n)}\right)\right) = \rho_{N,a}\left(I_R\left(N(n)\right)\right),$$

where we write $N(n)$ for $i^{(n)} = (i_1, \ldots, i_n)$ with $i_1 = \ldots = i_n = N$, $n \in \mathbb{N}_+$. From (4.3.2) we know that $I_R\left(i^{(n)}\right)$ is the interval with endpoints $\frac{p_n - p_{n-1}}{q_n - q_{n-1}}$ and $\frac{p_n}{q_n}$. Since $\frac{p_n}{q_n} = [i_1, \ldots, i_n]_R$, $n \in \mathbb{N}_+$, we get

$$\frac{p_n}{q_n} = \begin{cases} 1 - \dfrac{N}{i_1 + 1}, & \text{if } n = 1, \\ 1 - \dfrac{N}{i_1 + \dfrac{p_{n-1}(i_2, \ldots, i_n)}{q_{n-1}(i_2, \ldots, i_n)}}, & \text{if } n \geq 2, \end{cases}$$

and

$$\frac{p_n - p_{n-1}}{q_n - q_{n-1}} = \begin{cases} 1 - \dfrac{N}{i_1}, & \text{if } n = 1, \\ [i_1, \ldots, i_{n-1}, i_n - 1]_R, & \text{if } n \geq 2, \end{cases}$$

$$= \begin{cases} 1 - \dfrac{N}{i_1}, & \text{if } n = 1, \\ 1 - \dfrac{N}{i_1 + \dfrac{p_{n-1}(i_2, \ldots, i_n - 1)}{q_{n-1}(i_2, \ldots, i_n - 1)}}, & \text{if } n \geq 2. \end{cases}$$

Next, we can write

$$P_{N,i_1\ldots i_n}(a) = \frac{(a + N - 1)N^{n-1}}{(a + i_1)q_{n-1}(i_2, \ldots, i_n) - Nq_{n-2}(i_3, \ldots, i_{n-1}, i_n)} \quad (4.9.6)$$

$$\times \frac{1}{(a + i_1)q_{n-1}(i_2, \ldots, i_{n-1}, i_n - 1) - Nq_{n-2}(i_3, \ldots, i_{n-1}, i_n - 1)}$$

for all $i_n \in \mathbb{N}_N$, $n \geq 2$, and $a \in I$. Also by (4.9.6) we have

$$P_{N,N(n)}(a) = \frac{(a + N - 1)N^{n-1}}{(a + N)q_{n-1}(\underbrace{N, \ldots, N}_{(n-1) \text{ times}}) - Nq_{n-2}(\underbrace{N, \ldots, N}_{(n-2) \text{ times}})}$$

$$\times \frac{1}{(a + N)q_{n-1}(\underbrace{N, \ldots, N}_{n-2 \text{ times}}, N, N - 1) - Nq_{n-2}(\underbrace{N, \ldots, N}_{n-3 \text{ times}}, N - 1)}.$$

It is easy to see that $P_{N,N(n)}(\cdot)$ is a decreasing function. Therefore

$$\sup_{s \in I} \rho_{N,a}\left(s^a_{N,n} = s\right) = P_{N,N(n)}(a) \geq P_{N,N(n)}(1)$$

for all $a \in I$. □

Now, Proposition 4.9.1 allows us a very simple proof of a result which provides a lower bound for the supremum (4.9.1).

Proposition 4.9.2 (The Lower Bound [68]) *For any $a \in I$ we have*

$$\frac{1}{2} P_{N,N(n)}(1)$$

$$\leq \sup_{x,y \in I} \left| \rho_{N,a}\left(R^n_N \in [0,x], s^a_{N,n} \in [0,y]\right) - \frac{1}{\log\left(\frac{N}{N-1}\right)} \right.$$

$$\left. \times \log \frac{(x+N-1)(y+N-1)}{(N-1)(N-(1-x)(1-y))} \right|$$

for all $n \in \mathbb{N}_+$.

Proof Whatever $a \in I$ and $n \in \mathbb{N}_+$, by Proposition 4.9.1 we have

$$\sup_{x,y \in I} \left| \rho_{N,a}\left(R^n_N \in [0,x], s^a_{N,n} \in [0,y]\right) - \frac{1}{\log\left(\frac{N}{N-1}\right)} \right.$$

$$\left. \times \log \frac{(x+N-1)(y+N-1)}{(N-1)(N-(1-x)(1-y))} \right|$$

$$\geq \sup_{y \in I} \left| \rho_{N,a}\left(R^n_N \in [0,1], s^a_{N,n} \in [0,y]\right) - \frac{1}{\log\left(\frac{N}{N-1}\right)} \log\left(\frac{y+N-1}{N-1}\right) \right|$$

$$= \sup_{y \in I} \left| \rho_{N,a}\left(s^a_{N,n} \in [0,y]\right) - \rho_N([0,y]) \right| \geq \frac{1}{2} P_{N,N(n)}(1).$$

□

Remark 4.9.3 ([68]) Since $q_n(\underbrace{N, \ldots, N}_{(n-1) \text{ times}}, N-1) = N^n$, we get

4.9 A Two-Dimensional Gauss-Kuzmin Theorem

$$P_{N,N(n)}(1) = \frac{1}{q_n(N(n))}, \quad n \in \mathbb{N}_+.$$

By the recurrence relation (4.2.9) with $r_n = i_n$ for all $n \in \mathbb{N}_+$, we obtain

$$q_n(N(n)) = \frac{N^{n+1} - 1}{N - 1}.$$

It should be noted that Proposition 4.9.2 in connection with the limit

$$\lim_{n \to \infty} \left(\frac{1}{2} P_{N,N(n)}(1)\right)^{1/n} = \lim_{n \to \infty} \left(\frac{N-1}{2(N^{n+1}-1)}\right)^{\frac{1}{n}} = \frac{1}{N}$$

leads to an estimate of the order of magnitude of the error term $e_{N,n}(\rho_{n,a}; x, y)$.

In what follows we exploit the characteristic properties of the transition operator associated with the random system with complete connections underlying Rényi-type continued fraction. By restricting this operator to the Banach space of functions of bounded variation on I, we derive an explicit upper bound for the supremum (4.9.1).

Proposition 4.9.4 (The Upper Bound [68])
For any $a \in I$ we have

$$\sup_{x,y \in I} \left| \rho_{N,a}\left(R_N^n \in [0,x], s_{N,n}^a \in [0,y]\right) - \frac{1}{\log\left(\frac{N}{N-1}\right)} \right.$$

$$\left. \times \log \frac{(x+N-1)(y+N-1)}{(N-1)(N-(1-x)(1-y))} \right|$$

$$\leq \left(\frac{1}{N} + K_N\right)^n$$

for all $n \in \mathbb{N}$, where K_N is as in (4.5.10).

Proof Let $F_{N,n}^a(y) = \rho_{N,a}(s_{N,n}^a \leq y)$ and $H_{N,n}^a(y) = F_{N,n}^a(y) - \rho_N([0, y])$, $a, y \in I$, $n \in \mathbb{N}$. Note that $H_{N,n}^a(0) = 0$. As we have noted U_N is the transition operator of the Markov chain $(s_{N,n}^a)_{n \in \mathbb{N}}$. For any $y \in I$ consider the function f_y defined on I as

$$f_y(a) := \begin{cases} 1 & \text{if } 0 \leq a \leq y, \\ 0 & \text{if } y < a \leq 1. \end{cases}$$

Hence $U_N^n f_y(a) = E_{\rho_{N,a}}\left(f_y(s_{N,n}^a)\big| s_{N,0}^a = a\right) = \rho_{N,a}(s_{N,n}^a \leq y)$ for all $a, y \in I, n \in \mathbb{N}$.
As

$$U_N^\infty f_y = \int_0^1 f_y(a) d\rho_N(a) = \rho_N([0, y]), \quad y \in I.$$

It follows from Corollary 4.5.3 that

$$|H_{N,n}^a(y)| = \left|\rho_{N,a}(s_{N,n}^a \leq y) - \rho_N([0, y])\right| = \left|U_N^n f_y(t) - U_N^\infty f_y\right|$$
$$\leq \left(\frac{1}{N} + K_N\right)^n \operatorname{var} f_y = \left(\frac{1}{N} + K_N\right)^n \quad (4.9.7)$$

for all $a, y \in I, n \in \mathbb{N}$. By the very definition of the conditional probability and (4.4.15), for all $a \in I, x, y \in I$, and $n \in \mathbb{N}$ we have

$\rho_{N,a}\left(R_N^n \in [0, x], s_{N,n}^a \in [0, y]\right) = \rho_{N,a}\left(R_N^n \in [0, x] \big| s_{N,n}^a \in [0, y]\right)$
$\cdot \rho_{N,a}(s_{N,a} \in [0, y])$
$= \rho_{N,a}\left(R_N^n \in [0, x] \big| s_{N,n}^a \in [0, y]\right) \cdot F_{N,n}^a(y)$
$= \int_0^y \rho_{N,a}\left(R_N^n \in [0, x] \big| s_{N,n}^a = z\right) dF_{N,n}^a(z)$
$= \int_0^y \frac{Nx}{N - (1-x)(1-z)} dF_{N,n}^a(z) = \int_0^y \frac{Nx}{N - (1-x)(1-z)} d\rho_N(z)$
$+ \int_0^y \frac{Nx}{N - (1-x)(1-z)} dH_{N,n}^a(z) = \frac{1}{\log\left(\frac{N}{N-1}\right)} \log \frac{(x+N-1)(y+N-1)}{(N-1)(N-(1-x)(1-y))}$
$+ \frac{Nx}{N - (1-x)(1-z)} H_{N,n}^a(z)\Big|_0^y + \int_0^y \frac{Nx(1-x)}{(N-(1-x)(1-z))^2} H_{N,n}^a(z) dz.$

Hence, by (4.9.7)

$\left|\rho_{N,a}\left(R_N^n \in [0, x], s_{N,n}^a \in [0, y]\right) - \frac{1}{\log\left(\frac{N}{N-1}\right)} \log \frac{(x+N-1)(y+N-1)}{(N-1)(N-(1-x)(1-y))}\right|$
$\leq \left(\frac{1}{N} + K_N\right)^n \left(\frac{Nx}{N - (1-x)(1-y)} - \frac{Nx}{N - (1-x)(1-z)}\Big|_{z=0}^{z=y}\right)$
$= \left(\frac{1}{N} + K_N\right)^n \frac{Nx}{N-1+x} \leq \left(\frac{1}{N} + K_N\right)^n$

where K_N is as in (4.5.10), $a, x, y \in I, n \in \mathbb{N}$. $\square$

4.9 A Two-Dimensional Gauss-Kuzmin Theorem

Combining Proposition 4.9.2 with Proposition 4.9.4, we obtain Theorem 4.9.5.

Theorem 4.9.5 ([68]) *Whatever $a \in I$ we have*

$$\frac{1}{2} P_{N,N(n)}(1) \leq \sup_{x,y \in I} |\rho_{N,a}\left(R_N^n \in [0,x], s_{N,n}^a \in [0,y]\right)$$

$$- \frac{1}{\log\left(\frac{N}{N-1}\right)} \log \frac{(x+N-1)(y+N-1)}{(N-1)(N-(1-x)(1-y))}| \leq \left(\frac{1}{N} + K_N\right)^n$$

for all $n \in \mathbb{N}_+$.

Actually, Theorem 4.9.5 implies that the convergence rate is $\mathcal{O}(\delta_N^n)$, with

$$\frac{1}{N} \leq \delta_N \leq \frac{1}{N} + \frac{2}{2N - 1 + 2\sqrt{N(N-1)}}.$$

For example, we have (Table 4.4).

Also, the graph bellow suggests that for very large values of N the lower and upper bounds are very close. The graphs for the lower and upper bounds were obtained from the functions $f_{lb} : \mathbb{N}_+ \to \mathbb{R}$, $f_{lb}(N) = \frac{1}{N}$ and, respectively, $f_{ub} : \mathbb{N}_+ \to \mathbb{R}$, $f_{ub}(N) = \frac{1}{N} + \frac{2}{2N-1+2\sqrt{N(N-1)}}$ (Fig. 4.2).

Table 4.4 Lower and upper bounds for some $N \geq 2$

$N = 2$	$0.5 \leq \delta_N \leq 0.843145\ldots$
$N = 3$	$0.33333\ldots \leq \delta_N \leq 0.535374\ldots$
$N = 5$	$0.2 \leq \delta_N \leq 0.311456\ldots$
$N = 10$	$0.1 \leq \delta_N \leq 0.152668\ldots$
$N = 100$	$0.01 \leq \delta_N \leq 0.0150252\ldots$
$N = 1000$	$0.001 \leq \delta_N \leq 0.00150025\ldots$
$N = 10000$	$0.0001 \leq \delta_N \leq 0.000150003\ldots$

Fig. 4.2 Graphs of lower and upper bounds

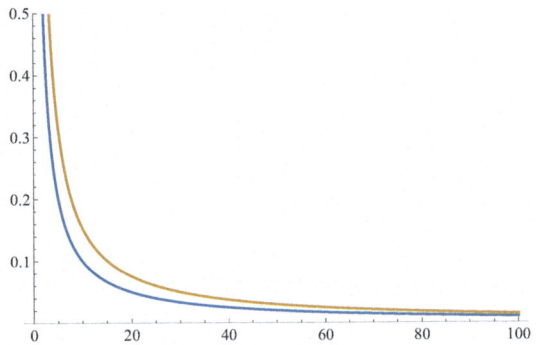

Comparing the Efficiency of Different Continued Fraction Algorithms

5

Since there are several continued fraction algorithms, we ask ourselves which of them provides the best approximation of a real number. The representation of a real number by a continued fraction can be viewed as a source of information about the number. For this purpose we need the notion of entropy, a rigorous tool at the crossroads between probability, information theory, and dynamical systems.

5.1 Entropy of Dynamical Systems

As is well-known, entropy is an important concept of information in physics, chemistry, and information theory [52]. The connection between entropy and the transmission of information was first studied by C. Shannon in [71]. Thus, the entropy can be seen as a measure of randomness of the system, or the average information acquired under a single application of the underlying map. Entropy also plays an important role in ergodic theory. Thus Shannon's probabilistic notion of entropy was first introduced into the ergodic theory by A.N. Kolmogorov [36] via a measure-theoretic approach. The contribution of Kolmogorov to modern dynamics was the discovery of the concept of entropy, which was made rigorous by Ya.G. Sinai [72]. This concept provides an important generalization of Shannon entropy. Kolmogorov-Sinai (K-S) entropy measures the maximal loss of information for the iteration of finite partitions in a measure preserving transformation. The concept has shown its strength through the adequate answers to problems in the classification of dynamical systems. Two metrically isomorphic dynamical systems have the same K-S entropy, so this concept is a tool for distinguishing non-isomorphic (nonconjugate) dynamical systems.

We briefly review this very important concept of K-S entropy in Ergodic Theory. Given a measure preserving system $(X, \mathcal{X}, \mu, T)$, we say $\alpha = \{A_i : i \in \Lambda\}$ is a *partition* of X

if $X = \bigcup_{i \in \Lambda} A_i$, where $A_i \in \mathcal{X}$ for each $i \in \Lambda$ and $A_i \cap A_j = \emptyset$ for $i \neq j$, $i, j \in \Lambda$. Here Λ is a finite or countable index set. For a partition α of X, we define *the entropy of the partition α* as

$$H(\alpha) := -\sum_{A \in \alpha} \mu(A) \log \mu(A). \tag{5.1.1}$$

In this definition T does not appear. However, the entropy of the dynamical system is defined by the entropy of the transformation T as follows.

Given a partition $\alpha = \{A_i : i \in \Lambda\}$ of X, then $T^{-k}\alpha = \{T^{-k}A_i : i \in \Lambda\}$, $k \in \mathbb{N}$, is as well. Given two partitions $\alpha = \{A_i : i \in \Lambda_1\}$ and $\beta = \{B_j : j \in \Lambda_2\}$, their least common refinement is $\alpha \vee \beta := \{A_i \cap B_j : i \in \Lambda_1, j \in \Lambda_2\}$. We are now in a position to state the definition of the entropy of transformation T. Thus, given a partition α, we consider the partition

$$\alpha_n := \bigvee_{i=0}^{n-1} T^{-i}\alpha = \left\{\bigcap_{i=0}^{n-1} T^{-i} A_i : A_i \in \alpha, i = 0, 1, \ldots, n-1\right\}. \tag{5.1.2}$$

Then, *the entropy of transformation T with respect to α* is given by

$$h(\alpha, T) := \lim_{n \to \infty} \frac{1}{n} H(\alpha_n). \tag{5.1.3}$$

The entropy of T is defined as

$$h(T) := \sup\{h(\alpha, T) : \alpha \text{ partition of } X \text{ such that } H(\alpha) < \infty\}. \tag{5.1.4}$$

In general, it does not seem possible to calculate the entropy straight from its definition. First, let us define that a partition α of X is a *generator* with respect to a non-invertible transformation T if the smallest σ-algebra containing all $T^{-k}A_i$, $k \in \mathbb{N}$, $i \in \Lambda$, is $\mathcal{X}$, i.e.,

$$\sigma\left(\bigvee_{i=0}^{\infty} T^{-i}\alpha\right) = \mathcal{X} \text{ up to sets of } \mu\text{-measure zero}. \tag{5.1.5}$$

In computation of $h(T)$ the Kolmogorov-Sinai Theorem is very useful. For completeness, we recall this theorem. Let α be a partition of X such that $H(\alpha) < \infty$. If α is a generator with respect to T, then $h(T) = h(\alpha, T)$.

We also have the following classical Shannon-McMillan-Breiman Theorem [14]. Let $(X, \mathcal{X}, \mu, T)$ be an ergodic measure preserving system and let α be a finite or countable partition of X satisfying $H(\alpha) < \infty$. The Shannon-McMillan-Breiman theorem says if $A_n(x)$ denotes the unique element $A_n \in \alpha_n$ such that $x \in A_n$, then for almost every $x \in X$

5.2 A Lochs-Type Approach

we have

$$\lim_{n\to\infty} -\frac{1}{n} \log \mu(A_n(x)) = h(\alpha, T). \qquad (5.1.6)$$

In 1964 V.A. Rohlin [54] showed that, when Rényi's condition is satisfied the entropy of a μ-measure preserving operator $T : X \to X$ is given by the formula:

$$h(T) := \int_X \log |T'(x)| \, d\mu(x). \qquad (5.1.7)$$

The *Rényi's condition* [52] means that there is a constant $C \geq 1$ such that

$$\sup_{x,y \in T^n(\alpha_n)} \frac{|u'_n(x)|}{|u'_n(y)|} \leq C, \qquad (5.1.8)$$

where $u_n := \left(T^n|_{\alpha_n}\right)^{-1}$.

5.2 A Lochs-Type Approach

The purpose of this section is to compare the efficiency of some continued fractions in approximating a real number in the unit interval. For any irrational $x \in (0, 1)$ suppose we have two known expansions $x = [a_1, a_2, \ldots]_1$ and $x = [b_1, b_2, \ldots]_2$. A natural question is: which of these continued fraction expansions is more "efficient"? In mathematical terms "efficiency" means which of the two sequences $[a_1, a_2, \ldots, a_n]_1$, respectively $[b_1, b_2, \ldots, b_n]_2$, converges faster to x as $n \to \infty$? It is therefore relevant to ask how much information (e.g., in terms of the digits in the second expansion) can be determined once we know n digits of the first expansion. Suppose we approximate x by keeping the first n digits of its first expansion. In order to attain the same degree of accuracy, we need to keep the first $m = m(n, x)$ digits of the second expansion. What can we say about the ratio $m(n, x)/n$ in general? The answer is that the relative speed of approximation of two different expansions (almost everywhere) is related to the quotient of the entropies of the transformations that generate these expansions.

Expansions that furnish increasingly good approximations to real numbers are usually related to dynamical systems. Although comparing dynamical systems seems difficult, using detailed knowledge of the continued fraction operator, G. Lochs [45] was able to relate the relative speed of approximation of decimal and regular continued fraction expansions (almost everywhere) to the quotient of the entropies of their dynamical systems.

Any real number $x \in [0, 1)$ can be written as

$$x = \sum_{i=1}^{\infty} \frac{d_i(x)}{10^i}, \tag{5.2.1}$$

where $d_i = d_i(x) \in \{0, 1, 2, \ldots, 9\}$ for $i \in \mathbb{N}_+$. The representation of x in (5.2.1), denoted by $x = 0.d_1 d_2 \ldots$, is called the *decimal expansion* of x. This expansion is generated by iterating the *decimal map*

$$T_d : [0, 1) \to [0, 1); \quad T_d(x) = 10x - \lfloor 10x \rfloor. \tag{5.2.2}$$

In other words, T_d is given by $T_d(x) = 10x - i$ if $\frac{i}{10} \leq x < \frac{i+1}{10}$, $i = 0, 1, 2, \ldots, 9$. Thus, we obtain

$$x = \frac{d_1}{10} + \frac{d_2}{10^2} + \ldots + \frac{d_n}{10^n} + \frac{T_d^n(x)}{10^n}, \tag{5.2.3}$$

where $d_1 = d_1(x) = \lfloor 10x \rfloor$ and $d_n = d_n(x) = d_1\left(T_d^{n-1}(x)\right)$, $n \geq 2$. Since $0 \leq T_d^n(x) < 1$, we obtain $\sum_{i=1}^{n} \frac{d_i}{10^i} \to x$ as $n \to \infty$.

Suppose that the irrational number $x \in (0, 1)$ has the decimal expansion $x = 0.d_1 d_2 \ldots$ and the RCF expansion $x = [a_1, a_2, \ldots]_G$ in (1.1.3) generated by iterating the Gauss map. Let $y_n = 0.d_1 d_2 \ldots d_n$ be the rational number determined by the first n decimal digits of x, and let $z_n = y_n + 10^{-n}$. Then, $[y_n, z_n]$ is the n-th order decimal interval containing x, which we also denote by $C_n(x)$. Now let $y_n = [b_1, \ldots, b_l]_G$ and $z_n = [c_1, c_2, \ldots, c_k]_G$ be the RCF expansions of y_n and z_n. Let

$$m(n, x) := \max\{i \leq \max(l, k) : \text{for all } j \leq i, b_j = c_j\}. \tag{5.2.4}$$

In other words, $m(n, x)$ is the largest integer such that $C_n(x) \subset D_{m(n,x)}(x)$, where $D_j(x)$ denotes the j-th order RCF interval containing x.

Lochs [45] proved that, for almost every irrational $x \in (0, 1)$, we have

$$\lim_{n \to \infty} \frac{m(n, x)}{n} = \frac{6 \log 2 \log 10}{\pi^2} = 0.97027014\ldots \tag{5.2.5}$$

Thus, roughly 97 RCF digits are determined by 100 decimal digits, which indicates that the RCF expansion is slightly more efficient compared to the decimal expansion at representing irrational numbers.

5.3 Dajani-Fieldsteel Theorem

K. Dajani and A. Fieldsteel [13] proved a generalization of Lochs' theorem showing that we can compare any two expansions of numbers which are generated by number-theoretic fibred maps, i.e., surjective interval maps $S : [0, 1) \to [0, 1)$ that satisfy the following conditions:

(1) There exists a finite or countable partition of $[0, 1)$ into intervals such that S restricted to each interval is strictly monotonic and continuous.
(2) S is ergodic with respect to Lebesgue measure λ, and there exists an S invariant probability measure μ equivalent to λ (i.e., $\mu(A) = 0$ if and only if $\lambda(A) = 0$ for all Lebesgue sets A) with bounded density.

Consider NTFMs S_1 and S_2 on $[0, 1)$ with invariant measures μ_1 and μ_2 (equivalent to Lebesgue measure) and with partitions P and Q, respectively. Denote by $P_n(x)$ the n-th order interval that contains $x \in [0, 1)$ (a similar definition for $Q_n(x)$). Let

$$m_{S_1, S_2}(n, x) = \sup\{m : P_n(x) \subset Q_m(x)\}. \tag{5.3.1}$$

Under the conditions just stated and with the understood that $m_{S_1, S_2}(n, x)$ is exactly the number of digits in the S_2-expansion of x that can be determined from knowing the first n digits in the S_1-expansion, we have

$$\lim_{n \to \infty} \frac{m_{S_1, S_2}(n, x)}{n} = \frac{h(S_1)}{h(S_2)} \quad \lambda - \text{a.e.}, \tag{5.3.2}$$

where $h(S_1)$ and $h(S_2)$ denote the entropy of S_1 and S_2, respectively, with $h(S_1) > 0$ and $h(S_2) > 0$.

5.4 Applications of the Rohlin's Entropy Formula

We study the families of continued fractions presented in the previous chapters, and in addition we consider the Chan's continued fractions studied by us in [38, 62]. To apply the result of Dajani and Fieldsteel in Sect. 5.3, we use the metric properties of these expansions. Applying Rohlin's formula, we compute the entropy of each map that generates these expansions.

Let $(X, \mathcal{X}, \mu, T)$ be the dynamical system associated with the continued fraction expansion whose digits a_k, $k \in \mathbb{N}_+$ take values in Λ. Let $\alpha = \{I(i)\}_{i \in \Lambda}$, where $I(i) = \{x \in \Omega : a_1(x) = i\}$ and Ω is the set of irrationals in $X \subset \mathbb{R}$. The partitions $\alpha_n := \bigvee_{i=0}^{n-1} T^{-i} \alpha$, $n \geq 1$, contain the n-th order fundamental intervals $I(i_1, \ldots, i_n) = \{x \in \Omega : a_1(x) = i_1, \ldots, a_n(x) = i_n\}$, $i_\ell \in \Lambda$, $\ell = \overline{1, n}$. By considering the functions

$u_i := \left(T|_{I(i)}\right)^{-1}$, $i \in \Lambda$, we have

$$u_{i_1...i_n} := u_{i_1} \circ u_{i_2} \circ \ldots \circ u_{i_n} = \left(T^n|_{I(i_1,...,i_n)}\right)^{-1}, \quad i_1, \ldots, i_n \in \Lambda.$$

5.4.1 θ-Continued Fraction Expansions

For a fixed irrational $\theta \in (0, 1)$, such that $\theta^2 = 1/s$, $s \in \mathbb{N}_+$, the transformation T_θ in (2.1.2) which generates the θ-expansion (2.1.1) is ergodic with respect to the invariant probability measure γ_θ in (2.1.3). For $(u_{\theta,i})_{i \geq s}$ defined in (2.4.1) we obtain

$$u_{\theta,b_1 b_2 \ldots b_n}(t) := (u_{\theta,b_1} \circ u_{\theta,b_2} \circ \ldots \circ u_{\theta,b_n})(t) = \cfrac{1}{b_1 \theta + \cfrac{1}{b_2 \theta + \cfrac{\cdot}{\cdot \cdot + \cfrac{1}{b_n \theta + t}}}}$$

(5.4.1)

and $u_{\theta,b_1 b_2 \ldots b_n} = \left(T_\theta^n|_{I_\theta(b_1, b_2, \ldots, b_n)}\right)^{-1}$, where $I_\theta(b_1, b_2, \ldots, b_n) = \{u_{\theta,b_1 \ldots b_n}(t) : t \in [0, \theta)\}$ is the n-th order fundamental interval. Such intervals with the endpoints $\frac{p_{\theta,n}}{q_{\theta,n}}$ and $\frac{p_{\theta,n} + \theta p_{\theta,n-1}}{q_{\theta,n} + \theta q_{\theta,n-1}}$, with $p_{\theta,n} := p_n$ and $q_{\theta,n} := q_n$ defined in (2.2.2) and (2.2.3) form a partition of $[0, \theta]$. Using some properties proved in Sect. 2.2, we check Rényi's condition

$$\frac{\left|u'_{\theta,b_1 \ldots b_n}(t)\right|}{\left|u'_{\theta,b_1 \ldots b_n}(r)\right|} = \left(\frac{q_{\theta,n} + r q_{\theta,n-1}}{q_{\theta,n} + t q_{\theta,n-1}}\right)^2 \leq \left(\frac{q_{\theta,n} + \theta q_{\theta,n-1}}{q_{\theta,n}}\right)^2 \leq \left(1 + \theta^2\right)^2$$

and we compute the entropy

$$h(T_\theta) = \int_0^\theta \log \left|T'_\theta(x)\right| d\gamma_\theta(x) = \int_0^\theta \frac{-\log x^2}{\log(1 + \theta^2)} \frac{\theta \, dx}{1 + \theta x}$$

$$= \frac{-2\theta}{\log(1 + \theta^2)} \int_0^\theta \frac{\log x}{1 + \theta x} dx.$$

(5.4.2)

5.4.2 Chan's Continued Fraction Expansions

H.C. Chan (see [11, 12]) considered a family of continued fraction expansions of any real number in the unit interval related to random Fibonacci-type sequences, whose digits are differences of consecutive nonpositive integer powers of an integer $\ell \geq 2$. The case $\ell = 2$ was thoroughly investigated in [61] and [30].

5.4 Applications of the Rohlin's Entropy Formula

Fix an integer $\ell \geq 2$. Then, any $x \in [0, 1)$ can be written in the form

$$x = \cfrac{\ell^{-c_1(x)}}{1 + \cfrac{(\ell-1)\ell^{-c_2(x)}}{1 + \cfrac{(\ell-1)\ell^{-c_3(x)}}{1 + \cdots}}} =: [c_1(x), c_2(x), c_3(x), \ldots]_\ell. \quad (5.4.3)$$

Define integer-valued functions $p_{\ell,n}(x) = p_n(x)$ and $q_{\ell,n}(x) = q_n(x)$ by

$$p_n(x) := \ell^{c_n(x)} p_{n-1}(x) + (\ell-1)\ell^{c_{n-1}(x)} p_{n-2}(x), \quad n \geq 2 \quad (5.4.4)$$

$$q_n(x) := \ell^{c_n(x)} q_{n-1}(x) + (\ell-1)\ell^{c_{n-1}(x)} q_{n-2}(x), \quad n \geq 1, \quad (5.4.5)$$

with $p_0(x) = 1$, $q_0(x) = 1$, $p_1(x) = 1$, $q_{-1}(x) = 0$, and $c_0(x) \equiv 0$.
Chan proved that

$$\frac{p_{\ell,n}(x)}{q_{\ell,n}(x)} := [c_1(x), c_2(x), \ldots, c_n(x)]_\ell \to x, \quad n \to \infty. \quad (5.4.6)$$

This continued fraction is associated with the following transformation τ_ℓ on I:

$$\tau_\ell(x) := \begin{cases} \dfrac{\ell^{\frac{\log x^{-1}}{\log \ell} - \left\lfloor \frac{\log x^{-1}}{\log \ell} \right\rfloor} - 1}{\ell - 1} & \text{if } x \neq 0 \\ 0 & \text{if } x = 0. \end{cases} \quad (5.4.7)$$

We notice that τ_ℓ maps the set of irrationals in I into itself. For any $x \in (0, 1)$, put

$$c_n = c_n(x) := c_1\left(\tau_\ell^{n-1}(x)\right), \quad n \in \mathbb{N}_+, \quad (5.4.8)$$

with $\tau_\ell^0(x) = x$ and

$$c_1 = c_1(x) := \begin{cases} \lfloor \log x^{-1} / \log \ell \rfloor & \text{if } x \neq 0 \\ \infty & \text{if } x = 0. \end{cases} \quad (5.4.9)$$

The transformation τ_ℓ which generates the continued fraction expansion (5.4.3) is ergodic with respect to an invariant probability measure, γ_ℓ, where

$$\gamma_\ell(A) := k_\ell \int_A \frac{dx}{\{(\ell-1)x + 1\}\{(\ell-1)x + \ell\}}, \quad A \in \mathcal{B}_I, \quad (5.4.10)$$

with

$$k_\ell := \frac{(\ell-1)^2}{\log\{\ell^2/(2\ell-1)\}}. \tag{5.4.11}$$

Define $(u_{\ell,i})_{i\in\mathbb{N}}$ by

$$u_{\ell,i}: I \to I; \quad u_{\ell,i}(x) := \frac{\ell^{-i}}{1+(\ell-1)x}. \tag{5.4.12}$$

For each $i \in \mathbb{N}$, $u_{\ell,i} = \left(\tau_\ell|_{I_\ell(c_1)}\right)^{-1}$. Let

$$u_{\ell,c_1c_2\ldots c_n}(t) := \left(u_{\ell,c_1} \circ u_{\ell,c_2} \circ \ldots \circ u_{\ell,c_n}\right)(t) = \cfrac{\ell^{-c_1(x)}}{1+\cfrac{(\ell-1)\ell^{-c_2(x)}}{1+\cdots+\cfrac{(\ell-1)\ell^{-c_n}}{1+(\ell-1)t}}}. \tag{5.4.13}$$

We obtain $u_{\ell,c_1c_2\ldots c_n} = \left(\tau_\ell^n|_{I_\ell(c_1,c_2,\ldots,c_n)}\right)^{-1}$, where

$$I_\ell(c_1, c_2, \ldots, c_n) = \{u_{\ell,c_1\ldots c_n}(t) : t \in [0,1)\} \tag{5.4.14}$$

is the n-th order fundamental interval. Such intervals with endpoints $\frac{p_{\ell,n}}{q_{\ell,n}}$ and $\frac{p_{\ell,n}+(\ell-1)\ell^{c_n}p_{\ell,n-1}}{q_{\ell,n}+(\ell-1)\ell^{c_n}q_{\ell,n-1}}$ form a partition of I.

Before applying Rohlin's formula, we must check Rényi condition (5.1.8). We use directly, without mentioning them here, some properties proved in [38]. Thus, we have

$$\frac{|u'_{\ell,c_1\ldots c_n}(t)|}{|u'_{\ell,c_1\ldots c_n}(r)|} = \left(\frac{q_{\ell,n}+r(\ell-1)\ell^{c_n}q_{\ell,n-1}}{q_{\ell,n}+t(\ell-1)\ell^{c_n}q_{\ell,n-1}}\right)^2 \leq \left(\frac{q_{\ell,n}+(\ell-1)\ell^{c_n}q_{\ell,n-1}}{q_{\ell,n}}\right)^2 \leq \ell^2. \tag{5.4.15}$$

Applying Rohlin's formula (5.1.7), we obtain

$$\begin{aligned}
h(\tau_\ell) &= \int_0^1 \log|\tau_\ell'(x)|\,d\gamma_\ell = \int_0^1 \log\left(\frac{\ell^{-c_1(x)}}{(\ell-1)x^2}\right) d\gamma_\ell \\
&= \int_0^1 (2\log(1/x) - c_1(x)\log\ell - \log(\ell-1))\,d\gamma_\ell \\
&= 2k_\ell \int_0^1 \frac{\log(1/x)}{((\ell-1)x+1)((\ell-1)x+\ell)}\,dx \\
&\quad - k_\ell(\log\ell)\int_0^1 \frac{c_1(x)}{((\ell-1)x+1)((\ell-1)x+\ell)}\,dx - \log(\ell-1).
\end{aligned} \tag{5.4.16}$$

5.4 Applications of the Rohlin's Entropy Formula

5.4.3 N-Continued Fraction Expansions

Fix an integer $N \geq 1$. The transformation T_N in (3.1.2) which generates the N-continued fraction expansion (3.1.1) is ergodic with respect to the invariant probability measure G_N in (3.1.3).

For $(v_{N,i})_{i \geq N}$ defined in (3.3.1) we obtain

$$v_{N,\varepsilon_1\varepsilon_2\ldots\varepsilon_n}(t) := \left(v_{N,\varepsilon_1} \circ v_{N,\varepsilon_2} \circ \ldots \circ v_{N,\varepsilon_n}\right)(t) = \cfrac{N}{\varepsilon_1 + \cfrac{N}{\varepsilon_2 + \cfrac{\cdots}{} + \cfrac{N}{\varepsilon_n + t}}} \quad (5.4.17)$$

and $v_{N,\varepsilon_1\varepsilon_2\ldots\varepsilon_n} = \left(T_N^n\big|_{I_N(\varepsilon_1,\varepsilon_2,\ldots,\varepsilon_n)}\right)^{-1}$. The intervals $I_N(\varepsilon_1, \varepsilon_2, \ldots, \varepsilon_n)$ with the endpoints $\frac{p_{N,n}}{q_{N,n}}$ and $\frac{p_{N,n}+p_{N,n-1}}{q_{N,n}+q_{N,n-1}}$ form a partition of I. Using some properties of $p_{N,n}$ and $q_{N,n}$ proved in [40], we verify Rényi's condition

$$\frac{\left|v'_{N,\varepsilon_1\ldots\varepsilon_n}(t)\right|}{\left|v'_{N,\varepsilon_1\ldots\varepsilon_n}(r)\right|} = \left(\frac{q_{N,n} + rq_{N,n-1}}{q_{N,n} + tq_{N,n-1}}\right)^2 \leq \left(\frac{q_{N,n} + q_{N,n-1}}{q_{N,n}}\right)^2 \leq \left(\frac{N+1}{N}\right)^2, \quad (5.4.18)$$

and we compute the entropy

$$h(T_N) = \int_0^1 \log\left|T_N'(x)\right| dG_N(x) = \frac{1}{\log\frac{N+1}{N}} \int_0^1 \frac{\log\left(\frac{N}{x^2}\right)}{x+N} dx$$

$$= \frac{\frac{\pi^2}{3} + 2\text{Li}_2(N+1) + \log(N+1)\log N}{\log\frac{N+1}{N}}, \quad (5.4.19)$$

where Li_2 denotes the *dilogarithm function*, defined by

$$\text{Li}_2(x) = \int_x^0 \frac{\ln(1-t)}{t} dt = \int_1^{1-x} \frac{\ln t}{1-t} dt \quad (5.4.20)$$

or

$$\text{Li}_2(x) = \sum_{k \geq 1} \frac{x^k}{k^2}. \quad (5.4.21)$$

5.4.4 Rényi-Type Continued Fraction Expansions

Fix an integer $N \geq 2$. The transformation R_N (4.2.1) which generates the Rényi-type continued fraction (4.2.5) is ergodic with respect to the invariant probability measure ρ_N in (4.2.6). For $(u_{N,i})_{i \geq N}$ defined in (4.4.1) we obtain

$$u_{N,r_1 r_2 \ldots r_n}(t) := \left(u_{N,r_1} \circ u_{N,r_2} \circ \ldots \circ u_{N,r_n}\right)(t) = 1 - \cfrac{N}{1 + r_1 - \cfrac{N}{1 + r_2 - \cfrac{\ddots}{\ddots - \cfrac{N}{r_n + t}}}}$$

and $u_{N,r_1 r_2 \ldots r_n} = \left(R_N^n \big|_{I_R(r_1,r_2,\ldots,r_n)} \right)^{-1}$, where $I_R(r_1, r_2, \ldots, r_n) = \{u_{N,r_1 \ldots r_n}(t) : t \in I\}$ is the n-th order fundamental interval. Such intervals with the endpoints $\frac{p_{N,n} - p_{N,n-1}}{q_{N,n} - q_{N,n-1}}$, $\frac{p_{N,n}}{q_{N,n}}$ with $p_{N,n} := p_n$ and $q_{N,n} := q_n$ defined in (4.2.8) and (4.2.9) form a partition of I. Using some properties proved in Sect. 4.2, we verify Rényi's condition

$$\frac{\left|u'_{N,r_1 \ldots r_n}(t)\right|}{\left|u'_{N,r_1 \ldots r_n}(r)\right|} = \left(\frac{q_{N,n} + (t-1)q_{N,n-1}}{q_{N,n} + (t-1)q_{N,n-1}}\right)^2 \leq \left(\frac{q_{N,n}}{q_{N,n} - q_{N,n-1}}\right)^2 \leq \left(\frac{N}{N-1}\right)^2$$

and we compute the entropy

$$h(R_N) = \int_0^1 \log \left|R'_N(x)\right| \, d\rho_N(x) = \frac{1}{\log \frac{N}{N-1}} \int_0^1 \frac{\log \frac{N}{(1-x)^2}}{x + N - 1} dx$$

$$= \log N + \frac{2\mathrm{Li}_2\left(\frac{1}{N}\right)}{\log \frac{N}{N-1}}, \qquad (5.4.22)$$

where Li_2 is as in (5.4.20). As examples, we have the values of $h(\tau_\ell)$, $h(T_\theta)$, $h(T_N)$ and $h(R_N)$ for different values of the parameters involved in Table 5.1.

Table 5.1 Some values of entropy for the transformations τ_ℓ, T_θ, T_N and R_N

ℓ	$h(\tau_\ell)$	s	$h(T_\theta)$	N	$h(T_N)$	N	$h(R_N)$
$\ell = 2$	1.62258	$s = 1$	2.37314	$N = 1$	2.37314	$N = 2$	2.37314
$\ell = 3$	1.26775	$s = 3$	3.24705	$N = 3$	3.24705	$N = 3$	2.905
$\ell = 5$	0.996315	$s = 5$	3.70244	$N = 5$	3.70244	$N = 5$	3.50063
$\ell = 10$	0.765943	$s = 10$	4.35074	$N = 10$	4.35074	$N = 10$	4.25052
$\ell = 50$	0.476521	$s = 50$	5.92195	$N = 50$	5.92195	$N = 50$	5.90194
$\ell = 100$	0.406218	$s = 100$	6.61015	$N = 100$	6.61015	$N = 100$	6.60015
$\ell = 200$	0.350849	$s = 1000$	8.90825	$N = 1000$	8.90825	$N = 1000$	8.90726

5.5 Comparing the Efficiency of Some Expansions

Now we focus on the investigation of the efficiency of several types of continued fraction expansions of a number in the unit interval. Thus, we aim to compare the efficiency by describing the rate at which the digits of one number-theoretic expansion determine those of another [44]. We apply the Dajani-Fieldsteel theorem presented in Sect. 5.3 and compare two by two the expansions discussed in the previous section. First we observe that for various values of the parameters involved, the entropies $h(T_\theta)$ and $h(T_N)$ are equal. Since entropy is an isomorphism invariant, we conjecture the following result.

Conjecture 5.5.1 For an irrational $\theta \in (0, 1)$ and a nonnegative integer $N \geq 2$ with $1/\theta^2 = N$, the transformations T_θ in (2.1.2) and T_N in (3.1.2) are isomorphic.

For this reason, we make only the following pairs: N-continued fractions and Chan's continued fractions, N-continued fractions and Rényi-type continued fractions, Rényi-type continued fractions and Chan's continued fractions. We observe that the transformations T_N, τ_ℓ, and R_N satisfy the two conditions from Dajani-Fieldsteel theorem.

5.5.1 N-Continued Fractions and Chan's Continued Fractions

Let $I_N^x(\varepsilon_1, \varepsilon_2, \ldots, \varepsilon_n)$ denote the n-th order interval of the N-continued fraction that contains x, and $I_\ell^x(c_1, c_2, \ldots, c_m)$ denote the m-th order interval of the Chan's continued fraction that contains x. Then

$$m_{N\ell}(n, x) := \sup \{m : I_N^x(\varepsilon_1, \varepsilon_2, \ldots, \varepsilon_n) \subset I_\ell^x(c_1, c_2, \ldots, c_m)\} \tag{5.5.1}$$

represents the number of digits in the Chan's continued fraction expansion of x in (5.4.3) that can be determined from knowing the first n digits in the N-continued fraction in (3.1.1). Therefore, applying (5.3.2) we have

$$\lim_{n \to \infty} \frac{m_{N\ell}(n, x)}{n} = \frac{h(T_N)}{h(\tau_\ell)}, \tag{5.5.2}$$

where $h(T_N)$ and $h(\tau_\ell)$ are as in (5.4.19) and (5.4.16), respectively. Given the values in Table 5.1, we observe that N-continued fraction expansion is more effective than Chan's continued fraction expansion regardless of the values taken by the parameters N and ℓ, respectively. As examples, we have

$$\lim_{n \to \infty} \frac{m_{12}(n, x)}{n} = 1.462571953\ldots \quad \text{or} \quad \lim_{n \to \infty} \frac{m_{32}(n, x)}{n} = 2.001164812\ldots \tag{5.5.3}$$

So roughly, if we approximate a number from the unit interval by keeping the first 1000 digits of the N-continued fraction expansion, in order to retain the same degree of accuracy we need to keep about 1462 digits in the Chan's continued fraction expansion.

5.5.2 N-Continued Fractions and Rényi-Type Continued Fractions

Let $I_N^x(\varepsilon_1, \varepsilon_2, \ldots, \varepsilon_n)$ denote the n-th order interval of the N-continued fraction that contains x, and $I_R^x(r_1, r_2, \ldots, r_m)$ denote the m-th order interval of the Rényi-type continued fraction that contains x. Then

$$m_{NR}(n, x) := \sup \left\{ m : I_N^x(\varepsilon_1, \varepsilon_2, \ldots, \varepsilon_n) \subset I_R^x(r_1, r_2, \ldots, r_m) \right\} \qquad (5.5.4)$$

represents the number of digits in the Rényi-type continued fraction of x in (4.2.5) that can be determined from knowing the first n digits in the N-continued fraction in (3.1.1). Therefore, applying (5.3.2) we have

$$\lim_{n \to \infty} \frac{m_{NR}(n, x)}{n} = \frac{h(T_N)}{h(R_N)}, \qquad (5.5.5)$$

where $h(T_N)$ and $h(R_N)$ are as in (5.4.19) and (5.4.22), respectively. Given the values in Table 5.1, we observe that N-continued fraction expansion is more effective than Rényi-type continued fraction expansion regardless of the values taken by the parameter N. We notice that $h(T_1) = h(R_2)$. We also have

$$\lim_{n \to \infty} \frac{m_{33}(n, x)}{n} = 1.117745267\ldots \quad \text{or} \quad \lim_{n \to \infty} \frac{m_{55}(n, x)}{n} = 1.057649623\ldots$$
$$\qquad (5.5.6)$$

As N grows, the entropies are very close, which means that the efficiency of the two continued fraction expansions are about the same.

5.5.3 Rényi-Type Continued Fractions and Chan's Continued Fractions

Let $I_R^x(r_1, r_2, \ldots, r_n)$ denote the n-th order interval of the Rényi-type continued fraction that contains x, and $I_\ell^x(c_1, c_2, \ldots, c_m)$ denote the m-th order interval of the Chan's continued fraction that contains x. Then

$$m_{R\ell}(n, x) := \sup \left\{ m : I_R^x(r_1, r_2, \ldots, r_n) \subset I_\ell^x(c_1, c_2, \ldots, c_m) \right\} \qquad (5.5.7)$$

represents the number of digits in the Chan's continued fraction of x in (5.4.3) that can be determined from knowing the first n digits in the Rényi-type continued fraction in (4.2.5). Therefore, applying (5.3.2) we have

$$\lim_{n\to\infty} \frac{m_{R\ell}(n,x)}{n} = \frac{h(R_N)}{h(\tau_\ell)}, \qquad (5.5.8)$$

where $h(R_N)$ and $h(\tau_\ell)$ are as in (5.4.22) and (5.4.16), respectively. Given the values in Table 5.1, we observe that Rényi-type continued fraction expansion is more effective than Chan's continued fraction expansion regardless of the values taken by the parameters N and ℓ, respectively. As examples, we have

$$\lim_{n\to\infty} \frac{m_{22}(n,x)}{n} = 1.462571953\ldots \quad \text{or} \quad \lim_{n\to\infty} \frac{m_{23}(n,x)}{n} = 1.871930586\ldots$$
(5.5.9)

5.6 Final Remarks

To conclude, N-continued fractions are more efficient than Rényi-type continued fractions at representing a number in the unit interval. Since the entropies $h(T_N) \geq 2.37314$ for $N \geq 1$, $h(R_N) \geq 2.37314$ for $N \geq 2$, and $h(\tau) = \pi^2/(6\log 2) = 2.37314$, it follows that N-continued fractions and Rényi-type continued fractions are more efficient than regular continued fractions (RCFs). Since the entropy $h(\tau_\ell) \leq 1.62258$ for $\ell \geq 2$, it follows that RCFs are more efficient than Chan's continued fractions. Thus, N-continued fractions are the most efficient at representing a number in the unit interval, with a very close efficiency being Rényi-type continued fractions.

Bibliography

1. Adler, R., & Flatto, L. (1991). Geodesic flows, interval maps, and symbolic dynamics. *Bulletin of the American Mathematical Society, 25*, 229–334.
2. Babenko, K. I. (1978). On a problem of Gauss. *Soviet Doklady Mathematics, 19*, 136–140.
3. Barnsley, M. (1993). *Fractals everywhere* (2nd ed.). Academic Press.
4. Barnsley, M., Demko, S., Elton, J., & Gerinomo, J. (1988). Invariant measures for Markov processes arising from iterated function systems with place-dependent probabilities. *Annales de l'Institut Henri Poincaré, Probabilités et Statistiques, 24*(3), 367–394.
5. Barnsley, M., & Elton, J. (1988). A new class of Markov processes for image encoding. *Advances in Applied Probability, 20*, 14–32.
6. Boyarsky, A., & Góra, P. (1977). *Laws of chaos: Invariant measures and dynamical systems in one dimension.* Birkhäuser.
7. Burger, E. B., Gell-Redman, J., Kravitz, R., Walton, D., & Yates, N. (2008). Shrinking the period lengths of continued fractions while still capturing convergents. *Journal of Number Theory, 128*(1), 144–153.
8. Burton, R., Kraaikamp, C., & Schmidt, T. (2000). Natural extensions for the Rosen fractions. *Transactions of the American Mathematical Society, 352*, 1277–1298.
9. Chakraborty, P. S., & Dasgupta, A. (2004). Invariant measure and a limit theorem for some generalized Gauss maps. *Journal of Theoretical Probability, 17*(2), 387–401.
10. Chakraborty, S., & Rao, B. V. (2003). θ-expansions and the generalized Gauss map. In K. Athreya, M. Majumdar, M. Puri & E. Waymire (Eds.), *Probability, statistics, and their applications: Papers in honor of Rabi Bhattacharya*. Institute of Mathematical Statistics, Lecture Notes-Monograph Series (Vol. 41, pp. 49–64).
11. Chan, H.-C. (2004). A Gauss-Kuzmin-Lévy theorem for a certain continued fraction. *International Journal of Mathematics and Mathematical Sciences, 20*, 1067–1076.
12. Chan, H.-C. (2006). The asymptotic growth rate of random Fibonacci type sequences II. *Fibonacci Quarterly, 44*(1), 73–84.
13. Dajani, K., & Fieldsteel, A. (2001). Equipartition of interval partitions and an application to number theory. *Proceedings of the American Mathematical Society, 129*(12), 3453–3460.
14. Dajani, K., & Kraaikamp, C. (2002). *Ergodic theory of numbers*. The Carus mathematical monographs.
15. Dajani, K., & Kraaikamp, C. (1999). A Gauss-Kuzmin theorem for optimal continued fractions. *Transactions of the American Mathematical Society, 351*, 2055–2079.
16. Dajani, K., Kraaikamp, C., & Van der Wekken, N. (2013). Ergodicity of N-continued fraction expansions. *Journal of Number Theory, 133*(9), 3183–3204.

17. Davison, J. L. (1977). A series and its associated continued fraction. *Proceedings of the American Mathematical Society, 63*(1), 29–32.
18. Doeblin, W. (1940). Remarques sur la théorie métrique des fractions continues. Compositio Mathematica, 7, 353–371.
19. Doeblin, W., & Fortet, R. (1937). Sur des chaînes à liaisons complètes. *Bulletin de la Société Mathématique de France, 65*, 132–148.
20. Durrett, R. (2005). *Probability theory: Theory and examples* (3rd ed.). Thomson Brooks/Cole.
21. Galambos, J. (1972). The distribution of the largest coefficient in continued fraction expansions. *The Quarterly Journal of Mathematics, 23*(2), 147–151.
22. Gröchenig, K., & Haas, A. (1966). Backward continued fractions, Hecke groups and invariant measures for transformations of the interval. *Ergodic Theory and Dynamical Systems, 16*, 1241–1274.
23. Haas, A. (2002). Invariant measures and natural extensions. *Canadian Mathematical Bulletin, 45*, 97–108.
24. Haas, A., & Molnar, D. (2004). Metrical diophantine approximation for continued fraction like maps of the interval. *Transactions of the American Mathematical Society, 356*, 2851–2870.
25. Harris, T. E. (1955). On chains of infinite order. *Pacific Journal of Mathematics, 5*, 707–724.
26. Iosifescu, M. (1963). Random systems with complete connections with an arbitrary set of states. *Revue Roumaine de Mathematiques Pures et Appliquees, 8*, 611–645.
27. Iosifescu, M. (1997). On the Gauss-Kuzmin-Lévy theorem III. *Revue Roumaine de Mathematiques Pures et Appliquees, 42*(1), 71–88.
28. Iosifescu, M., & Grigorescu, S. (2009). *Dependence with complete connections and its applications*. Cambridge tracts in mathematics (2nd ed., Vol. 96). Cambridge University Press.
29. Iosifescu, M., & Kraaikamp, C. (2002). *Metrical theory of continued fractions*. Kluwer Academic Publishers.
30. Iosifescu, M., & Sebe, G. I. (2006). An exact convergence rate in a Gauss-Kuzmin-Lévy problem for some continued fraction expansion. In *Mathematical analysis and applications, 90-109, AIP Conf. Proc.* (Vol. 835). American Institute of Physics.
31. Iosifescu, M., & Sebe, G. I. (2013). On Gauss problem for the Lüroth expansion. *Indagationes Mathematicae (N.S.), 24*, 382–390.
32. Iosifescu, M., & Theodorescu, R. (1969). *Random processes and learning*. Springer.
33. Karlin, S. (1953). Some random walks arising in learning models I. *Pacific Journal of Mathematics, 3*(4), 725–756.
34. Keane, M. S., Bedford, T., & Series, C. (Eds.) (1991). *Ergodic theory, symbolic dynamics, and hyperbolic spaces*. Oxford University Press.
35. Khintchine, A. Ya. (1963). *Continued fractions*. P. Noordhoff. Translated by Peter Wynn.
36. Kolmogorov, A. N. (1958). A new metric invariant of transient dynamical systems and automorphisms in Lebesgue spaces. *Doklady Akademii Nauk SSSR (N.S.), 119*, 861–864.
37. Kuzmin, R. O. (1928). On a problem of Gauss. *Doklady Akademii Nauk SSSR Series A* (pp. 375–380). [Russian; French version in *Atti Congr. Internaz.Mat. (Bologna*, 1928), Tomo **VI** (1932) 83–89. Zanichelli, Bologna].
38. Lascu, D. (2013). On a Gauss-Kuzmin-type problem for a family of continued fraction expansions. *Journal of Number Theory, 133*(7), 2153–2181.
39. Lascu, D. (2017). Metric properties of N-continued fractions. *Mathematical Reports, 19*(69)(2), 165–181.
40. Lascu, D. (2016). Dependence with complete connections and the Gauss-Kuzmin theorem for N-continued fractions. *Journal of Mathematical Analysis and Applications, 444*(1), 610–623.
41. Lascu, D., & Nicolae, F. (2017). A Gauss-Kuzmin-type problem for θ-expansions. *Publicationes Mathematicae Debrecen, 91*(3–4), 281–295.

42. Lascu, D., & Sebe, G. I. (2020). A dependence with complete connections approach to generalized Rényi continued fractions. *Acta Mathematica Hungarica, 160*, 292–313.
43. Lascu, D., & Sebe, G. I. (2020). A Gauss-Kuzmin-Lévy theorem for Rényi-type continued fractions. *Acta Arithmetica, 193*, 283–292.
44. Lascu, D., & Sebe, G. I. (2021). A Lochs-type approach via entropy in comparing the efficiency of different continued fraction algorithms. *Mathematics, 9*(3), 255.
45. Lochs, G. (1964). Vergleich der Genauigkeit von Dezimalbruch und Kettenbruch. *Abhandlungen aus dem Mathematischen Seminar der Universität Hamburg, 27*, 142–144.
46. Lévy, P. (1929). Sur les lois de probabilité dont dépendent les quotients complets et incomplets d'une fraction continue. *Bulletin de la Société Mathématique de France, 57*, 178–194.
47. Ma, L., & Nair, R. (2017). Haas Molnar Continued Fractions and Metric Diophantine Approximation. *Proceedings of the Steklov Institute of Mathematics, 299*(1), 57–177.
48. Mauldin, R. D., & Urbański, M. (2003). The doubling property of conformal measures of infinite iterated function systems. *Journal of Number Theory, 102*(1), 23–40.
49. Nakada, H. (1981). Metrical theory for a class of continued fraction transformations and their natural extensions. *Tokyo Journal of Mathematics, 4*, 399–426.
50. Norman, E. (1972). *Markov processes and learning models*. Academic Press.
51. Onicescu, O., & Mihoc, Gh. (1935). Sur les chaînes de variables statistiques. *Bulletin of Mathematical Sciences, 59*, 174–192.
52. Pollicott, M., & Yuri, M. (1998). *Dynamical systems and ergodic theory*. Cambridge University Press.
53. Rényi, A. (1957). Valòs szàmok elöallitàsàra szölgàlò algoritmusokròl.. *M. T. A. Mat. ès Fiz. Oszt. Kžl., 7*, 265–293.
54. Rohlin, V. A. (1964). Exact endomorphisms of a Lebesgue space. *American Mathematical Society Translations: Series, 39*, 1–36.
55. Ruelle, D. (1994). *Dynamical zeta functions for piecewise monotone maps of the interval*. CRM Monograph series (Vol. 4). American Mathematical Society.
56. Ryll-Nardzewski, C. (1951). On the ergodic theorems II (Ergodic theory of continued fractions). *Studia Mathematica, 12*, 74–79.
57. Schweiger, F. (1995). *Ergodic theory of fibred systems and metric number theory*. Clarendon Press.
58. Sebe, G. I. (2000). A two-dimensional Gauss-Kuzmin theorem for singular continued fractions. *Indagationes Mathematicae (N.S.), 11*(4), 593–605.
59. Sebe, G. I. (2001). On convergence rate in the Gauss-Kuzmin problem for grotesque continued fractions. *Monatshefte für Mathematik, 133*, 241–254.
60. Sebe, G. I. (2002). A Gauss-Kuzmin theorem for the Rosen fractions. Journal of Théorie des Nombres de Bordeaux, 14(2), 667–682.
61. Sebe, G. I. (2005). A Wirsing-type approach to some continued fraction expansion. *International Journal of Mathematics and Mathematical Sciences, 12*, 1943–1950.
62. Sebe, G. I. (2010). Convergence rate for a continued fraction expansion related to Fibonacci type sequences. *Tokyo Journal of Mathematics, 33*(2), 487–497.
63. Sebe, G. I. (2017). A near-optimal solution to the Gauss–Kuzmin–Lévy problem for θ-expansions. *Journal of Number Theory, 171*, 43–55.
64. Sebe, G. I., & Lascu, D. (2014). A Gauss-Kuzmin theorem and related questions for θ-expansions. *Journal of Function Spaces, 2014*, 12 p.
65. Sebe, G. I., & Lascu, D. (2019). On convergence rate in the Gauss-Kuzmin problem for θ-expansions. *Journal of Number Theory, 195*, 51–71.
66. Sebe, G. I., & Lascu, D. (2020). A two-dimensional Gauss-Kuzmin theorem for N-continued fraction expansions. *Publicationes Mathematicae Debrecen, 96*(3–4), 291–314.

67. Sebe, G. I., & Lascu, D. (2020). Convergence rate for Rényi-type continued fraction expansions. *Periodica Mathematica Hungarica, 81*(2), 239–249
68. Sebe, G. I., & Lascu, D. (2022). Two asymptotic distributions related to Rényi-type continued fraction expansions. *Periodica Mathematica Hungarica, 85*(2), 380–398
69. Sebe, G. I., & Lascu, D. (2022). Some asymptotic results for the continued fraction expansions with odd partial quotients. *Turkish Journal of Mathematics, 46*(7), 3011–3024
70. Series, C. (1985). The modular surface and continued fractions. *Journal of the London Mathematical Society, 2*, 69–80.
71. Shannon, C. (1948). A mathematical theory of communication.. *Bell Labs Technical Journal, 27*, 379–423.
72. Sinai, Ya.G. (1959). On the notion of entropy of a dynamical system. *Doklady Russian Academy of Sciences, 124*, 768–771.
73. Szüsz, P. (1961). Über einen Kusminschen Satz. *Acta Mathematica Academiae Scientiarum Hungaricae, 12*, 447–453.
74. Viswanath, D. (1961). Random Fibonacci sequences and the number 1.13198824..... *Mathematics of Computation, 69*(231), 1131–1155.
75. Van der Wekken, C. D. (2011). Lost periodicity in N-continued fraction expansions. Bachelor thesis, Delft University of Technology, Delft, 2011. http://repository.tudelft.nl/view/ir/uuid:67317fff-f3e3-44e4-8e59-51e70782705e/
76. Wirsing, E. (1974). On the theorem of Gauss-Kuzmin-Lévy and a Frobenius-type theorem for function spaces. *Acta Arithmetica, 24*, 506–528.

The manufacturer's authorised representative in the EU is Springer Nature Customer Service Centre GmbH, Europaplatz 3, 69115 Heidelberg, Germany. If you have any concerns regarding our products, please contact ProductSafety@springernature.com

Printed and bound by CPI Group (UK) Ltd, Croydon, CR0 4YY
26/03/2026
02078943-0019